Windmills

Old wooden windmill near Silver City, New Mexico.

Windmills

And Pumps
of the
Southwest

By Dick Hays
with Bill Allen

Drawings by Gordon Gilpin

EAKIN PRESS ᴘ Fort Worth, Texas
www.EakinPress.com

TABLE OF CONTENTS

FOREWORD

I started at an early age as a cowboy. That was when a puncher's jobs were many and varied; up close to the top of the list in importance was making sure water was pumped for the livestock. For the most part, windmills did that job. Consequently, regardless of how much my legs were bowed, I had to know a little something about water wells and windmills. What I did know I had learned from various cowboys and from my personal experiences — both good and bad — with windmills.

One day, while following a bunch of cattle along, with my feet hanging down under a half-broke horse, the thought hit me like a four-forty lightning bolt that the only place I had ever signed a check was on the back, and that other than my bedroll and a change of clothes, I was settin' on everything I owned. Perhaps I needed a haircut or maybe my hat was too tight, but other than Will Rogers, I couldn't recall too many cowboys putting much of a scar on the world. I decided then and there that a change of occupation might be in order. Thus it was that I set out with a note at the American National Bank, the unsolicited assurance of its vice president that I was going straight uphill with a worn out well rig, and a determination to water the Southwest.

Along about this time southwestern ranchers decided to go from quantity to quality with their cattle. Cattlemen wanted cows that weighed more, and wanted their herds to have more conformity, exhibiting characteristics of better breeds, like the Hereford. To control breeding, ranchers fenced their land into smaller pastures, and to improve grazing they began to practice range rotation procedures. Before too long, the cows themselves had changed — they didn't range as far, especially in rough country, as they used to. All these changes brought about the need for more wells and pumping equipment, mostly windmills. In those days — the 1920s and 1930s — a well driller could move in, drill a well, set up a windmill and put it to pumping, settle the cost right on the spot using a stick to do his figuring in the mud-run, get paid, then move on to the next job. For the most part, ranchers in those days understood wells and windmills and the maintenance that just naturally went along with them.

Nowadays, things aren't quite so simple. Farmers and ranchers are faced with a shortage of hands, and the ones they do have not only draw high wages, but are specialists in things other than windmills. Then, too, there's the influx of suit and necktie men into the countryside and their dependence on private water systems. Changes such as these have brought about a need for somebody who is an installation and maintenance specialist — the professional windmiller.

Over the years, even the professional has had to change. Consider, for example, the coming of the Rural Electrification Administration. Years ago, electricity was distributed to remote ranches and farms, and the windmiller soon found that he had to learn how to install and maintain electric pumps, or else go out of business.

This book was written to help the professional windmiller as well as the rancher, farmer, or country gentlemen who isn't able to hire a professional. What I'd like to do is share whatever my years of experience and mistakes have taught me. One thing that I learned right off in school was that the world was two-thirds water and one-third land. The first thing that I learned as a windmiller was that that rule wasn't true at all: the world is one-third water and two-thirds rock, and most of that rock was in the Southwest, directly under the ground where I wanted to drill a water well. Another thing I found out was that windmills, pump experts, or anyone connected with water wells believe that if it's square you roll it and if it's round you pick it up and carry it. Never do anything the easy way. If you as a windmill owner manage to miss one rock, or carry something square just once, then my efforts to put what I know in print will have been worthwhile.

Every windmill and pump manufacturer has instructions on how to install their product. These directions are pretty good—up to a point. What this book will do is take you beyond that point, out of the technical world—that lacks experience—into the practical one. I'll go into more detail concerning installation and maintenance of windmills and electric pumps, and do it so you'll be able to tackle whatever is wrong with your mill or pump a lot easier than the technical book way. Believe me, there are usually easier ways to install or fix one of these devices.

While we're talking about technical things, let me tell you about a few of the items I've tried to standardize in the writing. One is the mill itself. There are a number of good windmills on the market, but here in the Southwest the Aermotor seems to be the most popular. Thus it is that this book is primarily concerned with that brand. Bear in mind, though, that what's said about Aermotor applies pretty well to all other brands, whether they are still being made or not.

To make things even simpler, I've chosen a standard for wells and windmills. First, assume that the well is dug, that it has water in it, that it's straight and cased regardless of the underground rock formation, and that there is a cemented casing at the top to guard against surface contamination. This well is assumed to be two hundred feet deep. Above it sits a thirty-three foot steel Aermotor tower topped by a ten foot mill. Down at the bottom of the well is the cylinder — 1-7/8 inch diameter is the size I've chosen as standard. This is the arrangement that will be used for discussions about wells and windmills. I didn't choose to use a standard electric pump because there are simply too many kinds. We will discuss each one at the proper time.

There's one last thing I need to point out. When you get ready to install a new mill, or have one repaired, go to your professional windmiller first — not the distributor. A windmiller is usually an agent for the distributor, and the prices you pay for parts or for a whole windmill or pump will be the same whether or not you go to the distributor. When you use your local windmiller, you'll get the benefit of his experience along with the price. And you'll help him out, too. Instead of the distributor getting the windmiller's commission, the latter gets it — that's the way it should be.

Acknowledgment needs to be made at this point to the courtesy of the Aermotor Division, Valley Pump Corporation, which kindly furnished the illustrations that have been identified in the text. Also, allow me to extend my thanks to Gordon Gilpin, Silver City, New Mexico, who furnished the excellent pen and ink drawings.

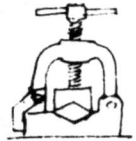

I

WHY A WINDMILL?

Granddad had a windmill,
and it went round and round
a squeakin' and a squawkin'
and pullin' water from th' ground.

Grandma raised a garden,
chickens, kids, and flowers,
with water from th' well
when there wasn't any showers.

Dad, he used that windmill too,
after Granddad passed away
and me, the oldest grandson
I'm still using it today.

One windmill, three generations,
I think that proves the worth
of a machine built by man
to draw water from th' earth.

In the Southwest, water sometimes spells the difference between life or death, profit or loss, win or lose. Not necessarily quality water: any water. Any water at all, just as long as it isn't too hard or too thick to run through a pipe.

There has never been much surface water in this arid country, including times gone by when the Indians, the Spaniards, and a few wagon trains, settlers, and prospectors were the only humans around. Streams and creeks abound, all right, but they're dry most of the year. For this reason those early occupants of our Southwest all followed routes of free flowing water; but these were few and far between. Imagine if you will a wagon train, dependent upon oxen that need water often, going a hundred miles out of its way, from water to water, when there was an abundance of it perhaps a hundred feet below, maybe even less. The point: there is water. I've often wondered how many battles were fought over a water hole, battles that needn't have been had those who wanted the water just tried to dig straight down instead of moving on to the next water that was a lot farther away — and in miles, not feet!

When the newcomers began to settle, their one vital concern was for permanent water. Ranchers were more or less limited in the amount of land they could use; it all depended on the location of the water. People dug wells, at first by and, and later, windmills came on the scene. There is no doubt that Winchester and Colt received a lot of credit for settling the Southwest that in reality belongs to the windmill.

The first windmills on this continent were probably water pumping devices in Mérida on the Yucatan Peninsula in the mid-1500s. As for North America, the French began to build windmills along the St. Lawrence River in 1629, and from then on their number increased rapidly in the East. Mostly these mills were used to grind grain, and they were custom made.

Later, factory-built machines began to make their appearance. Railroads used them to fill the tanks of thirsty steam engines, richer families bought them to provide fresh water, and ranchers and homesteaders in the western plains began to use them to water cattle.

In 1889, there were seventy-seven factories in the United States that made windmills, but by 1919 there were just thirty-one. One of the most celebrated was the self regulating "Halladay-type" whose wooden blades actually pivoted to parallel the wind before it was strong enough to damage the mill. Wheels of the Halladay ranged from eighteen to sixty feet in diameter: the Halladay Standard Windmill was manufactured until 1929. Another design, typified by the Eclipse solid-wheel windmill, worked very much like present day machines such as the Aermotor.

One historian confirms that in the windy West, besides the Colt revolver and barbed wire, cheap windmills were essential to survival on the Great Plains, just as they were later in the Southwest. A Nebraska child once wrote:

ndmills, not Winchester and Colt,
ttled the Southwest.

A tired old windmill in the Southwest.

> We like it in the sandhills
> We like it very good.
> For the wind it pumps our water
> And the cows they chop our wood.

Beginning in the 1870s metal bladed windmills began to replace those made of wood. There were a number of manufacturers of note. One was Sears, Roebuck & Company's Kenwood, whose tail was red bordered and whose sails were tipped in the same color. Among the oldest windmill companies still in existence today are the Dempster Engineering Works, the Fairbury Windmill Company, and the Heller-Aller Company that manufactures the Baker Direct Stroke Steel Mill. And, of course, there's the Aermotor Company, maker of a series of windmills that seems to be pretty popular here in the Southwest.

3

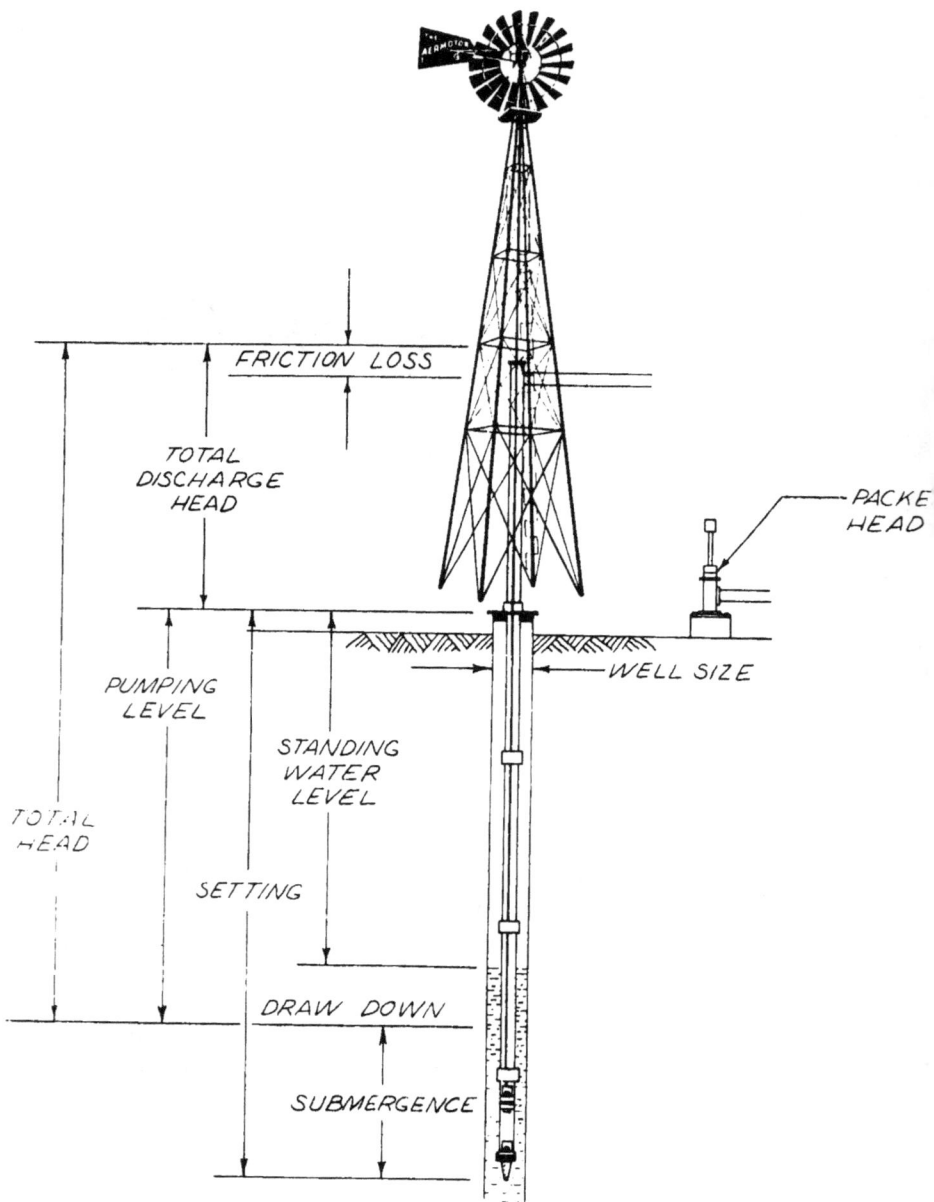

FRICTION LOSS

TOTAL
DISCHARGE
HEAD

PACKE
HEAD

PUMPING
LEVEL

WELL SIZE

STANDING
WATER
LEVEL

TOTAL
HEAD

SETTING

DRAW DOWN

SUBMERGENCE

The tail vane is all that's left of this old Dempster.

When the 1930s rolled around, there also came upon the scene the Rural Electrification Administration. Windmills, at least in places where cheap electricity became available, began to fall into disuse as electric pumps took their place.

Now, however, events are taking place that seem to be reversing that trend. Electricity is becoming more and more expensive, and there is a very real threat of shortage of fossil fuels that generate this power. Contrary to what many people might think, windmills are not a thing of the past. They never have been, and emphasis on their use is now being renewed. There are more windmills working, at least here in the Southwest, than ever before.

I guess before we go any further I should explain to you — at least basically — something about how a windmill works. The accompanying sketches have been simplified at this point: a little later on you'll see some drawings that show all the parts and just how each one works.

Windmills can have anywhere from sixteen to thirty-six metal sails (fan blades) forming a wheel that can range anywhere from six to sixty feet in diameter. This wheel is mounted on a strong hub that is in turn secured to a horizontal shaft. Affixed to the shaft are small gears that drive a set of larger gears. These large gears in turn actuate a Pitman arrangement which changes the circular motion of the shaft to a useable up and down stroke. All the gears and the Pitman are housed in a cast iron case — like an oil pan for an automobile engine —

5

Windmill personalities in the Southwest.

which contains the oil that constantly lubricates all the moving parts. The iron case is topped by a galvanized steel cover called, appropriately, the helmet. Attached to the case is a tail vane that keeps the wheel facing into the wind, much like a weather vane. The vane is mounted on a swivel so that it can manually be moved parallel to the wheel, causing the wheel to face away from the wind and effectively (in conjunction with a brake) turning the windmill off. Additionally, the tail vane is spring-loaded so that when the wind becomes too strong the pressure automatically forces the tail parallel to the wheel, shutting off the mill to keep it from running wild when the wind's blowing too hard.

Naturally, the entire mill assembly — wheel, case, and tail — has to be able to rotate freely about a vertical axis. This is done by placing it atop a rotating mast pole that in turn rests on the windmill tower. In this respect the Aeromotor mill is different than other brands; the mast pipe doesn't rotate. Instead, the mill assembly turns on a turntable. Towers, incidentally, can be anywhere from thirty to fifty feet high, depending on local wind conditions. Because winds are stronger as the distance above ground level increases, it's usually necessary to put a mill on a tower. Tower height can be critical. In any case, the tower must be high enough to clear a twenty-one foot length of pipe as well as the pipe-lifting equipment that normally hangs in the tower.

The Pitman mechanism is attached to a pump rod which in turn is connected to a sucker rod. The sucker rod — quite small in diameter — actually goes down to the bottom of the well through a water column pipe (often called the drop pipe, not to be confused with the well casing) and is connected to a plunger and an upper check valve. This plunger and valve are located inside a cylinder which is situated at the bottom of the drop pipe and which is smaller in diameter. Both plunger and upper check valve move in concert with the up and down motion — or stroke — of the windmill. In the bottom of the cylinder is another valve, known as the lower check valve or stationary valve. The plunger and two valves are what make the water go up the drop pipe by means of a lifting action. This is an important point: a windmill doesn't actually pump water, it lifts it up the water column pipe. This is how it's done. During the down stroke the lower check valve is forced closed and the upper check valve is forced open. What water that is between the two valves has no place to go except above the plunger, through the upper check valve. When the stroke is reversed, the lower check is opened, allowing water to be sucked into the void between plunger and lower check. Also during the up-stroke the upper valve is forced closed, and the plunger actually lifts a column of water a distance equal to the stroke of the windmill. It's this amount of water—whatever fits into a space the distance of the stroke by the diameter of the drop pipe — that is forced into a storage tank with each stroke of the wind-mill. Strokes of windmills can be changed. If you have a mill that needs to pump water from a shallow well then it should have a long stroke. But if your mill has to pump from a deep well it should have a shorter stroke. Naturally, it will pump less water in a given amount of time than a longer stroked mill.

Since I've already started to talk about how things work, it might be a good idea to mention briefly how pumps operate: they'll be covered in more detail later on. Basically, there are three kinds of pumps that ranchers, cowboys, windmillers, and gentlemen farmers might be dealing with.

The first of these devices is called a pump jack, which actually works on the same principle as the windmill. A pump jack is simply a motorized — electric or gasoline — substitute for the windmill itself. It's connected to the sucker rod and when it's operating it lifts water just like a windmill, using the same up and down stroke.

The second and most popular kind of pump is the submer-sible. Requiring electricity, the submersible pump, as its name implies, works in the water at the bottom of the well. There is no sucker rod. The pump is a cylindrical apparatus that screws on to the bottom of a discharge pipe and uses impellers driven by an electric motor to force the water up the pipe directly into a storage tank or into the home water system. A submersible pump is hooked to an electrical source by means of a water-proof conduit that goes from an above-ground control box to the pump motor in the bottom of the well. An important con-sideration here is that since you are using electricity where there's water, due care must be taken when working on sub-mersible pumping systems. Generally, submersibles are good choices for deep water wells.

One other kind of pump you might be dealing with is the jet

Pumpjacks old and new.

pump. The electric motor of the jet pump is above ground, which facilitates trouble-shooting. These pumps are often thought to be undependable, but they do work well if properly maintained. Jet pumps make use of a high-speed jet of water which is forced through a small pipe down the well to the level of the water. This jet of water is then directed back up the stand pipe, carrying with it part of the water in the well. As you can imagine, this pump has to be completely primed to operate: this is one of its problems. If a jet pump loses its prime it won't pump. Because this type of pump seems to be finicky, many people won't use it, but for certain applications they can't be beat. Like anything else, how well they work depends on how well they are maintained. Jet pumps are generally used only in shallow wells, but do have certain deep well applications.

★　★　★　★　★

So much for how things work. Let's talk for a while about what this chapter is supposed to be about: why a windmill?

The windmill, so far as I know, is the most economical and dependable means of pumping water in remote areas where electricity is not available. In such places water is essential to the survival of cattle. If you give a hungry, thirsty cow some food, some water, and some salt, she'll go to the water first — every time. I think I'd be safe in saying that nine out of ten windmills used today in the Southwest are for pumping water for livestock, usually in more remote locations.

The other one of the ten windmills may have been put up for several reasons. People often use them to provide water for secluded summer cabins. Or someone's neighbor may or may not have a windmill, commonly called keeping up with or getting ahead of the Joneses. Some windmills are installed, even though electricity's available, to provide water for permanent residences. One reason for this is that the owners feel that the wind is free, that electricity is expensive, and that fuel is in short supply, a problem that becomes more acute with each passing year. Often there are people with money to burn who just plain like windmills, who simply want to watch the wheel go around — the romantics. I've tried never to discourage anyone, whatever their reason for wanting a mill, although even with present day prices it's still cheaper to pump water with electricity than with a windmill. Windmills are not inexpensive to buy — or install.

A small Samson windmill provides water for an adobe church.

An 18-footer.

Here are a couple of tales — both tall — about the why of windmills. Naturally, they're not the best of reasons for getting yourself a windmill.

Once, a long time ago, I heard an old cowboy tell a dude who got off the EP&SW train at Hachita, New Mexico, that windmills were used in the Southwest as cow fans. That's not true.

Along that same railroad line, but a little farther west, there are a number of windmills with wheels that may range up to twenty feet in diameter. One day I asked a local old timer why the windmills were so big. It just didn't seem logical to me to have such large wheels, since the water isn't very far below ground in this particular part of New Mexico. (Deep wells require a big mill. Here, in most cases, a ten-foot mill would have done the job.) Now, what that fellow told me could or couldn't be true, but this is what he said. "When folks was a pourin' into this country just after the turn of the century, the ranchers that had spreads along the track wasn't about to invite any homesteaders in to stay. They did everything they could to keep 'em on the train, headed farther west to Arizona or California. One thing the ranchers did was to put up big mills, the biggest they could find. When the settlers looked out the train

11

window and saw those big mills a turnin' slowly in the breeze they said to themselves that water was too deep here for any farmin', and that they'd just as well head on to better places. That was just what the ranchers wanted, and it must have worked, because there never were too many of those homesteaders around these parts."

All I can say about those two reasons for wanting a windmill is that they don't seem too sound. Especially these days with all the educated dudes!

Regardless of your reason for putting up a windmill, the fact is that when they're properly installed and maintained, they are dependable, long-lived machines. Naturally, there are advantages and disadvantages, as there are with anything, in owning a windmill. I'd like to mention a few of these to get you on the right track in your thinking in case you are considering getting one — either new or used.

The advantages of a windmill are pretty obvious, especially here in the West where there's lots of space, few people, and sometimes no roads. Windmills work well in these remote areas of no electricity and generally pretty strong, consistent winds. The only other ways to water cattle in this kind of country are to build dirt tanks to catch runoff from the rains — not the most dependable water supply in the world — or to use gasoline engines to run pump jacks or power a generator that in turn runs a submersible pump. (More on this last method later, because it has certain advantages.) Consider, though, the pump jack and the engine hooked up to it: no matter how remote the well, sooner or later someone is going to walk off with the jack or the engine or both of them! It's a bit tougher to steal a windmill, although I know for a fact that this has been done, too. I heard recently that in certain parts of West Texas there were actually organized windmill "rustlers," who make a healthy profit when they move into an area. With the proper equipment they can take down a remote windmill in a matter of hours, then sell it somewhere else at a price well below the cost of a new one.

Another good thing about a windmill is that it doesn't have to be looked after all the time. It's dependable. Other than giving it adequate maintenance and a weekly check to make sure it's pumping water, you don't have to worry about it too much. It'll probably still be working after you're gone.

Another advantage may become more important as time goes on, as the price of electricity goes up, and as fossil fuels become more scarce. After the original expense of putting up a windmill, operating costs are almost nil. It may well be that someday windmills might be cheaper in the long run than installing an electric pump. I'd guess that this state of affairs is still a few years off, but nonetheless possible. When you figure that some windmills have been pumping for better than forty years, there's no question that sooner or later they'll pay for themselves.

There is one other plus for the windmill over the electric pump. Since every windmill has to have a companion storage tank of some kind, and since when they're pumping properly they can keep that tank full, you will usually have a supply of water in reserve. This isn't always true with an electric system. When you do have a windmill failure, there will be a few days leeway in getting it fixed because there is no immediate and complete loss of water.

What about the disadvantages of a windmill for water? In most cases they're the opposite of the advantages. Because they're entirely dependent on the wind, you have to provide some sort of water storage facility, whether it's a metal tank, a rim, or simply a hole scooped in the ground. Naturally, a windmill pumps a much lower volume of water than an electric pump, another reason for needing adequate storage. And, as I've mentioned before, windmills are initially expensive to install, much more so than electrical pumps and their associated controls. Also, you do need to maintain a windmill: the oil has to be changed yearly and all the bolts have to be tightened at the same interval. One big problem for some is that all this maintenance has to be done while you're perched on a tower at least thirty feet off of the ground. This factor alone contributes greatly to the difficulty of maintaining a windmill. And the maintenance required will vary with the kind of water that's being pumped. Some water is hard on the cylinder leathers, other water is easier; you might have to change them every year or every five years, depending. One last problem with windmills is that they're often the target of vandals or hunters. Somehow, to a few individuals, it seems to be good sport to perforate the tail vane, helmet, or sails of a windmill with the trusty .22 or 30-30. And tourists often drop things down the

13

stand pipe (as do pack rats), and what kid can resist climbing a windmill tower? Any or all of these efforts on the part of our fellow humans — and animals — can ruin a windmill, and do it quickly.

The decision as to whether you should or shouldn't have a windmill for water has to be your own. It depends on your circumstances and feelings about the situation. A big part of the decision has to be based on money. Count yours before you decide.

II

THE PROFESSIONAL WINDMILLER

The professional windmiller can often fix a windmill that's not pumping by hitting it with a hammer. He won't charge you a cent for hitting the mill. But he will charge you for knowing where to hit it!

I guess there are a lot of definitions for a windmiller, and I guess that I've heard most of them. Here's kind of a composite.

The professional is someone who has been horse-kicked in the head. He's usually a male from fifteen to seventy-five years old. His weight might vary from between ninety to two-hundred pounds, depending on his height and the amount of grease, rust, and mud accumulated on his person. He may be called by several different names: among those printable are windmiller, tower-monkey, Don Quixote, joint-buster, wind-jockey, or whatever. On rare occasions he may even be called by his name, but he should never be referred to as mister, since such practice might cause his head to swell until his hat won't fit. His diet should be, for the most part, coconuts and bananas. He may not gain any weight but he sure will be able to climb a windmill tower. A windmiller's reason for choosing this profession may have been varied, but I would say that probably first and most important was to keep his belt buckle from rubbing his spine. He is not especially noted for excessive wealth: the tip a windmiller leaves for the waitress may include two or three pieces of pocket change, a couple of flat washers, and a half-hearted smile. His clothes and general appearance are such that if he tarries too long in a town where he's not known, he may have to answer to a charge of vagrancy. While

Wherever there are windmills, there's a need for a professional.

others of more noble profession may at times perspire, the windmiller just plain sweats. Most likely he won't offer much competition for any man-of-the-year award.

Laying all the aforementioned malarky aside, a windmiller's services are as important to the community as any. A hundred or more head of cattle milling around and hooking each other away from a water trough where yellow jackets have drunk what water was left presents a problem that needs immediate attention. This is no place for a forty-hour a week person.

16

The Professional Windmiller

The telephone rang in th' middle of the night
nearly jumpin' off the wall,
the windmiller's wife rose from her bed
and sleepily answered th' call
The rancher's voice came loud and clear
from a hundred miles away
"I'm sorry if I got you out of bed
but please hear what I have to say."
"Oh! That's all right," th' good wife said
through a big broad sleepy yawn.
"I would have answered sooner
but I was out a'mowing the lawn."
Th' rancher went on with his tale of woe
an' these are the words he said:
"My cattle are starvin' for water
on th' south end of my spread.
Last week, th' windmill it quit pumpin'
and it makes me sad to tell,
me and th' wetbacks fixed it
but we dropped th' pipe in th' well.
Would you send your husband, with fishing tools,
as soon as you can get him up.
Now, I'm not one to rush him
I reckon yesterday might be soon enough.

Is there a need for the professional? I'll not attempt to dispute the fact that most any good, hard-working handyman or do-it-yourselfer can choose, buy, and install a windmill so that it will pump water for what seems like forever. The only difference between the windmiller and that kind of person is, for the most part, a matter of the range of tools owned and the years of experience possessed. About all else that the professional needs to know is how to tie a good double half hitch — with cable, rope, or chain — and be able to open and close a wire gate at least twenty times a day. (This brings up a problem facing the professional. Often he has to go through remote gates, sometimes locked. Here he uses one of the tools that he has, the good old acetylene torch. Instead of driving for hours trying to find the rancher, who probably lost the key anyway, the professional will simply cut the chain when he goes in to the pasture, and reweld it when he goes out. This is assuming the rancher has kept his bill current. If not, windmillers have been known to cut the lock, which can't be rewelded, and simply reweld the gate shut permanently, without benefit of a removeable lock.)

I have assumed throughout this book that the professional who reads it already knows quite a bit about windmills and pumps. In this and the following chapters the windmiller may find tips that he hasn't used before. If so, well and good, I'm glad that he learned something. By and large this chapter is devoted to business techniques and tools that may not have been thought of or used. If the professional needs to learn about windmill repairs he'll find it in subsequent chapters. Keep in mind that the professional will work on a lot of wells, topped by windmills or using electric pumps. For this reason, I will list and describe at least some of his techniques, tools and their uses, stressing multiple uses where possible.

Let's talk about advertising first. There is none better than the word-of-mouth kind that comes automatically with prompt and efficient service. When something comes up with service already performed, then quick no-cost resolution of the problem is important. Don't forget to do it cheerfully! This is real advertising. But for the beginner, you'll need more, and the best and simplest are the yellow pages with a good comprehensive ad. Don't forget newspapers, TV and radio, and local magazines: they are especially good when you're getting started. A magnetic or other weatherproof sign attached to your trucks will help, and a long-lasting sticker with your firm's name, address, and telephone number, that can be placed on a windmill tower leg or a pump control box, is good for repeat business. It is also a good idea to set up display windmills on high and low towers. They always catch the public's eye.

What service in the windmill business means is a full-time operation, on call seven days each week and twenty-four hours each day. When do you have a holiday? When no one is out of water!

This kind of service also means that there has to be someone available to answer the telephone all the time during those twenty-four hour seven-day weeks. Who does this? The windmiller's wife can do it — if she's so inclined — whenever there's no one at the shop. If he's a bachelor windmiller or one with a working gal, then he should hire the services of a very reliable answering service. Don't forget to follow up quickly on calls received when you're out of the office. Promptness, in this regard, often will spell the difference between success and failure of your windmill and pump business.

Another thing that I've found necessary to a good operation is to keep a good supply of the most commonly used parts in the windmill/pump shop or yard. Besides, in these days of high inflation, it's like money in the bank. Your supply would include such windmill-related items as pipe of all necessary dimensions; sucker rods, both metal and wood; pipe fittings such as unions and tees; cylinders of the most common varieties used in your area; and cup leathers of all sizes. Items for the pump business should include spare submersible and jet pump motors, impellers, submersible wire, and appropriate electrical supplies.

Whether your business is predominantly concerned with windmills, or most of your work is with pumps, there is no substitute for good records. High on the list of necessary bookwork is a good set of well logs. They are indispensible to efficient operation, and useless if not kept up to date.

My well logs consisted of 4x6 cards — one for each well — filed alphabetically by the owner's name. A sample card is shown below. Take a good look at it.

Mangas Cattle Co.   Box 00, Tyrone, N.M.
                        Blackhawk No. 1
Depth 200'   200'6" Casing   SWL 170'
Drilled & Cased 4-15-68   Driller - Joe Blow
4-20-68 — Installed 10'6O2 Aermotor - 33' Tower
          9 Jts 21'-189'2" R&D pipe - 1-1/8 ash rod - 1-7/8x30 ABBV cyl.
4-10-69 — Serviced
5-10-70 — Serviced
4- 1-71 — Serviced - releathered
4-20-72 — Serviced - replaced valves with O'Bannon cyl.

All the information you can get about the well should be recorded on the card. Such items as depth, casing size, and static water level (SWL) are important should you ever have, for example, to replace a windmill with a pump. When cards are kept up to date, you can actually do your figuring as to what will be needed in the way of a pump before you go to the well site. At least one trip is saved, maybe even more. Note that this sample indicates exactly what is installed in the way of a mill, water column pipe (nine joints of 21 foot reamed and drifted [R&D] 2 inch pipe for a total of 189 feet), sucker rod, and the cylinder. (Note that initially in 1968 a 1-7/8 inch diameter by 30 inch length all brass ball valve (ABBV) cylinder was installed, but that in 1972 an O'Bannon cylinder [a much more efficient kind] replaced it during routine servicing). Let's assume that you get a call in the middle of the night to go out and fix this rig, and that the mill is at least forty miles from nowhere. What do you do? Extract the appropriate well log card, throw whatever spares are necessary to fix the mill or cylinder in your truck, and go. No need to drive forever only to have to return to get whatever part or parts you needed. I've seen the time in the Southwest where the trip to one mill might take a half a day. Unnecessary trips like that tend to cut pretty deep into your profits! It goes without saying — almost — that you should never slight the job of keeping these cards current, and that your first duty when you get back to the shop is to update the appropriate cards. For some strange reason, parts for one kind of cylinder don't seem to fit other brands, and the wrong part at a remote mill site is as bad as than no part at all.

Another advantage of well logs is that they give you a good idea of when the mill needs to be serviced. This example shows routine servicing every year, and indicates that it took three years for the leathers to wear out on the cylinder. If the cylinder hadn't been replaced with an O'Bannon variety, which has no leathers, then the windmiller would have known that he ought to replace the leathers in 1974, and can prepare for this eventuality when the time comes. Why wait till the leathers wear out at an inopportune time? Preventive maintenance is the answer, and it will save you and your customers both time and money.

Ranchers, especially those with large operations and many windmills and pumps, should also keep a good set of well logs. The reasons are the same as for the professional. It simply makes the operation and maintenance of the mills much easier. The rancher will be able to check his own maintenance — when was the last time the one in the horse pasture was lubricated; when were the bolts on the tower last tightened; or when was the pump checked for efficient operation? Naturally, owner maintenance items such as these should be a part of the rancher's logs. Include whatever is important to you.

A double half hitch.

This is probably a good time to mention the help needed by a professional windmiller. This person ought to be someone who when he gets up in the morning and crosses his legs, doesn't take an hour and thirty minutes to figure out how to put his shoes on. Or put another way someone with common sense and no fear, but a lot of respect, for the type of work he has to do. A helper or two in the windmill and pump business is a necessity. Of primary importance is that your helpers know how to use tools; not just use them, but use them well. Just like the professional himself, the helper must, and I mean must, know how to tie a double half hitch with chain, cable, or rope. If he doesn't know, and can't be taught, then either get another helper or more insurance. Once I sent a helper with a windmill down US180 through a small settlement. I followed about thirty minutes later and found that the feathers on the chickens were still turned the wrong way. When I finally caught up with him, I had to explain that the 180 sign was the highway number, not the speed limit. Helpers like this you don't need!

It's a fact that a windmiller needs a lot of tools. Mostly they're expensive, so if you can use one tool to do a number of jobs, so much the better, especially for the pocket book. The ones that I'm going to tell you about are those that I've found to be most useful over the many years that I was in the windmill and pump business. Of course, there are a lot more you can buy — mostly gadgets — but they don't help that much. The ones I recommend are basic and important — without them the job will be much more difficult, if not impossible.

The biggest, and best and most universal tool — also by far the most expensive — is the pipe pulling unit. There are several good hydraulic pulling rigs on the market these days. My experience has been with a Model 33 Hydra-hoist manufactured by Baker Manufacturing Company. This I mounted on a one-ton pickup truck, which was also fitted with a standard service body and a pipe rack.

Its hoist is powered by a hydraulic pump driven by a power takeoff from the truck transmission. The pump in turn drives a hydraulic motor mounted on the hoist. Four three-position valves operate all segments of the hoist. There is one for each of two hydraulic jack pads, one for raising and lowering the mast, and one for the operation of the winch through a three-speed transmission which is mounted nearby. All segments are powered in both directions and are not dependent on any kind of braking system: this is one of the best features of this rig because it is completely safe. The only way you can drop anything with this machine is by breaking the cable. If a hydraulic line breaks, whatever is moving will stop because it is powered both ways.

The telescoping mast may be extended by raising the top section at two foot intervals by the winch to a length of thirty-two feet. Another ten-foot section may be bolted on the end for additional length, but this is not suitable for heavy loads.

There is an adjustable divider on the winch drum which separates the storage and working side of the drum. The working line should be 7/16 inch non-rotating cable with a good swivel hook equipped with a safety latch. (Spend your money on non-rotating cable, the hook, and the safety latch, because they really are important for safe operation.) At least three wraps of the cable should be left on the working side of the drum divider. When the swivel hook is extended to the load area the rest of the cable is spooled on the storage side of the drum. For running a bailer or sand pump or when more cable is needed, the working line may be pulled off, and up to about eight-hundred feet of 3/8 inch sand line can be spooled by

A Hydra-hoist.

The simple Hydra-hoist
controls.

The winch drum.

A rig can be set up in five minutes.

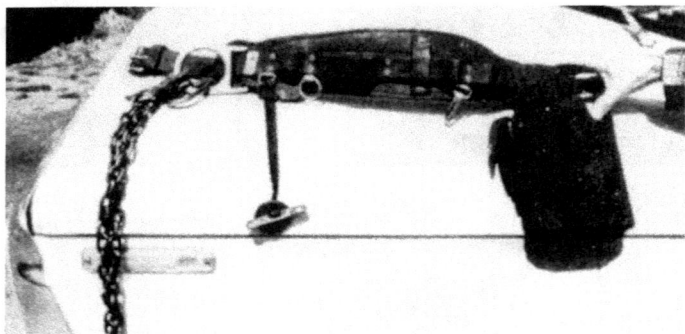

A good safety belt is essential.

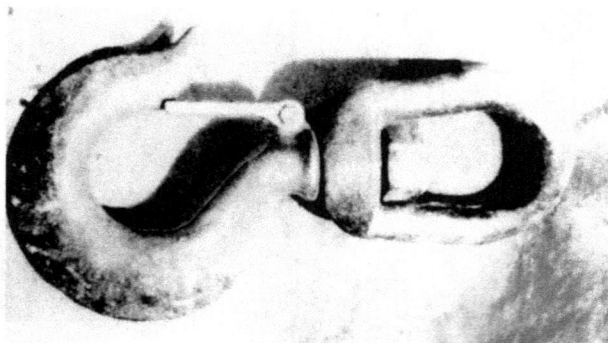

Swivel hook with safety latch.

sliding the divider all the way to the left side of the drum.

The load capacity of the Hydra-hoist is six thousand pounds, more by doubling the load line. This rig is big enough to handle most ordinary well jobs and still small enough to get into the many tight places in the Southwest. A good operator can set this rig up and be ready to start working in less than five minutes.

I can sum up my feelings about the Hydra-hoist or an equivalent rig in one sentence: I'm sure there are men who have made a success of the windmill and pump business without a piece of equipment such as this, but I don't know how they did it!

Now for some of the lesser pieces of equipment. Let's talk first about safety belts. At least two good strong belts with some tool loops should be kept with your rig and yet in a protected place where they will not become damaged. That's important, because the wearer's life may well depend on them. One man with two free hands is worth more than three men hanging on and working — almost handicapped — with one hand. Working on top of a windmill is no place for daredevils. A lineman's climbing belt, by replacing the strap that goes around the pole with a small strong chain and a double snap in the D-ring on the dead end for a quick adjustment, makes an ideal belt for the windmiller. Don't ever go to the field without it.

Another item of equipment that's indispensable is an acetylene torch with cutting and welding tips. This piece of equipment — which will have many uses — should be mounted on the rig truck. Besides cutting and welding, heating babbit material for bearings, heating pipe collars to loosen joints, starting a fire with wet wood for making coffee, with an enriched acetylene mixture it makes for an excellent night light when you have to finish a job after dark. And as I've already mentioned, the torch is great for cutting chains on locked gates in remote areas. Just be sure to weld it back when you leave. Perhaps, though, I'd better clarify that last operation a bit. That should only be done when the windmiller has been sent for and not provided with a key to whatever pasture he has to go through. Cutting gate chains just to be cutting is a good way to beat old age.

Pipe and sucker rod elevators — all expensive — come in various sizes and types and are used in pairs. Their purpose is to

A welding set mounted on the truck is handy.

A typical pipe elevator.

The specially cut boiler plate.

simply provide a means to hold the pipe or sucker rod when it is being raised or lowered in the well. Rather than buy several sets of costly elevators it is better to buy a larger size—usually for three or four inch pipe—then make bushings for each smaller pipe size that bolt into the elevator slots. The one thing of importance is that since the elevators may be required to lift an entire lengthy load of pipe with a single collar as a bearing and lifting surface the fit must be snug.

A piece of half-inch boiler plate, with a notch cut out that is large enough to clear the collars, should be kept with the elevators. This plate is placed on top of the casing. It is used as a support for the elevators when they are at rest just before a drop pipe is pulled up.

This seems like a good place to stress the importance of keeping the load line tight at all times. No slack, no kinks, and for sure, no sudden jerks. Up until the time of our space ventures it was generally assumed that what went up came down — Newton and apple, I think, were the ones that pointed that out. But it's also a well known fact that what goes down may not necessarily come up. If you should happen to drop a string of pipe or sucker rod in a well it may not be a major disaster for the local population, but the stunt does have its drawbacks. If no one was hurt you have already used up about half your luck. The other half will be needed in fishing for what was dropped down the well. Then you can always figure on a ruined cylinder and a few lengths of dog-legged pipe, depending on how much and how far it was dropped. Following is listed the complete set of instructions for what to do in the event you do manage to drop something big — like a string of pipe — down the well. the well.

RUN TILL TH' RACKET STOPS! Do not attempt to catch it.

Concerning wrenches, for most well work I don't recommend extremely large size pipe wrenches. For one reason, they are too awkward to handle, and for another, when only one wrench is used to loosen a tight pipe joint it may cause the pipe to be bulged inwards, what we call "egging" the pipe. Two men with the best kind of wrenches — 24-inch chain tongs — can usually loosen most pipe joints. If that doesn't work, simply

27

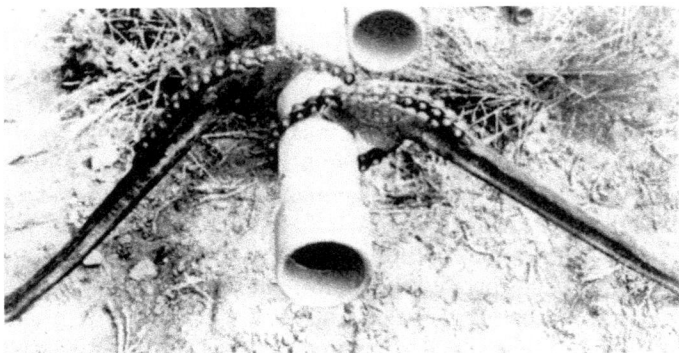

Here's how to use the chain tongs to hold pipe.

The extremely useful
chain tong.

Chain tongs have a
variety of uses.

A home-made slip-spear.   A home-made overshot.

use a cheater, a pipe that will slip over the handle and effectively lengthen it and increase the leverage. Another 24-inch chain tong, with an eye welded on to the handle so that you can tie it securely to a tower leg with a chain, makes a good solid back up arrangement to either eliminate a man or provide a solid base for pulling against. The teeth in these wrenches are designed so that by shoving the wrench forward you can turn the pipe in the opposite direction without removing the wrenches. These teeth are fastened securely to the wrench with a set-screw. When they become worn and dull they can easily be replaced with a new set of teeth. Also, a set of tongs fastened around a standpipe or a tower leg makes an excellent place to stand while working on something that would normally be out of reach. Two of these wrenches may be used to hold pipe while cutting and threading it when no vise is available.

In addition to the three chain tongs a couple of sixteen or eighteen-inch standard pipe wrenches should be in your tool inventory. They'll be used for small pipe and for sucker rod.

You'll always find a need for pipe cutters and dies. You should have two sizes of pipe cutters with you, one to cut one-eighth to two inch pipe, the other to cut two to four inch pipe. It's also a good idea to always keep a couple of spare cutting wheels in the rig. You'll also need pipe dies that will thread one-eighth to four inch pipe.

In case you or someone else (sometimes a rancher) has dropped pipe or sucker rod down the well, there are ways, for the most part unsatisfactory, of getting it out. These ways use things we call fishing tools, and they are usually home-made. You'll need overshots for collars for trying to catch the outside of dropped pipe or sucker rod, and you'll need slip-spears for trying to nab the inside of two to four-inch pipe.

Handy items to have with your rig are a couple of small come-alongs. They can be used for all kinds of jobs, such as binding loads on a truck or securing packages of pipe or sucker rod.

An assortment of good chain — and I stress the word good — is a must. Chains should range from five-sixteenth to three-eighth-inch in diameter and be anywhere from three to ten feet long. You should install clevis grab hooks on each end of every chain. Clevis grab hooks only because in case of some sort of hangup where you can't release the tension on the chain enough to remove the hook, you can simply knock the pin out of the hook, releasing it from the chain.

A windmill and pump man has to have a measuring wire and a weight. The cheapest and most accurate means of measuring the location of water in a well is with a small strong flexible wire rolled on a spool. Tie a weight of not less than five pounds to the end (an old sash weight does the job well). You'll need about a thousand feet of number nine (stay) wire, with no kinks, and the spool should be about six inches in diameter. Let the weight down the well, and when it hits bottom, mark the wire with a bit of electrician's tape. Slip the wire into a snatch block, and begin to pull it directly away from the well. Make note of where the water begins to appear on the sheave of the snatch block, and again mark the cable. When the wire is all the way out of the well use a hundred-foot steel tape to measure both the static water level and the depth of the well.

Anyone connected with water well work, at one time or another, is going to need to look down a well. When tourists stop by for a visit, they'll want to look, too. This is best accomplished by using sunlight and at least one, perhaps two, mirrors. You'd be surprised how bright that light is when reflected down to the bottom of a well. For a better view, use binoculars. (Binoculars are also handy for checking windmills from a long distance to see whether or not they are pumping.)

Another thing that is always needed at any mill site is a tail rope, more appropriately called a hand line. This rope should be at least a half-inch in diameter and long enough to reach double through a snatch block on the tail of the windmill. A five-gallon can attached to the rope is good for raising and lowering tools and for oil changes. (This same can, turned upside down at the well head, also makes a handy place to keep tools out of the mud or dust.) The two ends of the rope, when tied together, enable the operator to have control of whatever object he may be raising or lowering.

One good rule is not to stand directly under anything being hoisted up or down unless it's absolutely necessary, and then be sure and keep a good lookout. While we're on this subject, make it a rule never to stand at all under the man working on the tower, as you never know when he might drop something. As for the man on the tower, he should never drop tools and things from above: he should always use the tail rope. If only one man is on the job he can raise what he needs to the top, tie the rope to the tower, then climb up without the load. This rope will have a lot of other uses, but never attempt to hang a bird hunter with it.

This is as good a place as any to mention climbing a windmill tower by means of the ladder. Do not trust rungs as something to hold onto with your hands. Always take hold of ladder rails, or in the case of Aermotor towers, take hold of the tower leg. Rungs are for feet, not hands.

There's something about cans and drums that make them really handy for the windmiller. I couldn't begin to operate without a bunch of old coffee cans, handy for keeping just about everything in. As for old drums, they make excellent portable tables for tools at the well. Because they are so expensive these days, anything you can use to keep tools from getting lost is a plus. You'd be surprised how quickly a tool will disappear when you're pulling a three-inch pipe with enough water in it to turn the immediate area into a muddy

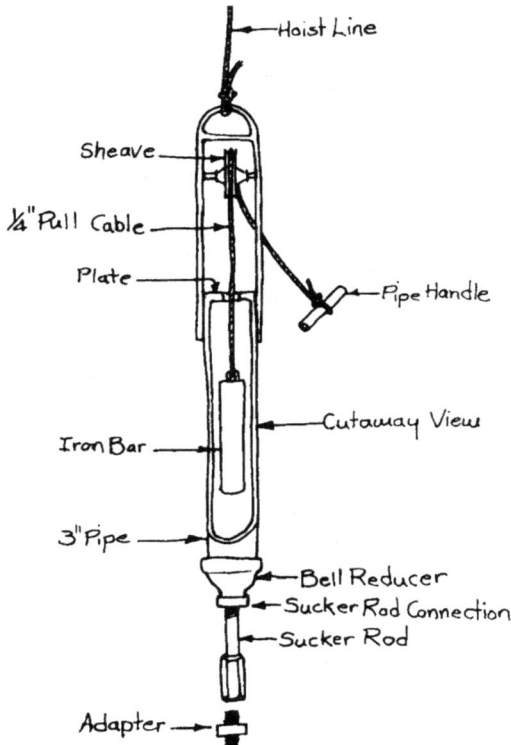

Labels on diagram:
- Hoist Line
- Sheave
- ¼" Pull Cable
- Plate
- Pipe Handle
- Cutaway View
- Iron Bar
- 3" Pipe
- Bell Reducer
- Sucker Rod Connection
- Sucker Rod
- Adapter

Rod Jar

mess. (One other thing you can do to lessen this possibility is to wrap burlap sacks around the joint that will soon be loosened — this will tend to keep the water from spraying all over just when you don't want it to.)

One other home-made tool is often necessary, and it's easily made. Many times after screwing onto the bottom check valve, you'll find it so tight that it will actually lift the entire water column. The diagram shows how to make a device called a jar that can be screwed on to the sucker rod connection and hooked to the load line which takes the weight of the load. The windmiller actually jerks the run cable, and that will usually loosen the valve. You should never use jacks to try to pull this valve, as you may end up pulling everything apart. I'll go into more detail about this whole procedure a little later when we're talking about maintenance.

For the windmiller, the last thing on the tool list is an assortment of hand tools that should always be kept right on the truck. The list should include a shovel; pick; crowbar; axe; four or five vise-grips; a hand brace and an assortment of both wood and steel drilling bits; a woodsaw; hacksaw and several extra blades; a sledge hammer; a claw hammer; a ball peen hammer; a set of square socket wrenches; channel lock pliers of several sizes; side cutting pliers; standard pliers; needle-nose pliers; long taper punches; drift punches; center punches; cold chisels; draw knife; various files; several

This huge Challenge is used at a ranch headquarters.

screwdrivers; a one-hundred-foot steel tape; an eight-foot pocket take; a level; and probably many more that I've failed to mention. The windmiller will usually end up with more paraphernalia than a new mother, and what he throws away today will probably be the first thing he will need tomorrow. One thing that will help, though, is keeping all this gear — at all times — in it's proper place on the truck. If you can't find it when you need it, you're wasting time — and money.

There are not too many more tools needed by the professional to expand his operation to the installation and repair of electric pumps. One of them is a combination volt and ammeter. You should also have an assortment of fuses, capacitors, overload relays for both 115 and 230 volt submersible pumps, pressure switches, pressure gauges, foot valves, and impellers for jet pumps.

Since most of his work will be with pumps instead of windmills, the professional should have in his shop — or have immediate access to — submersible pumps and spare motors from at least one-half to one-and-one-half horsepower. In most cases these pumps operate a pressure system and when they fail they need immediate attention because there is no reserve water supply. This means instant replacement. Depending entirely on a distant supplier is very poor business in this case.

33

III

HOW TO BUY A WINDMILL

A windmill is just a windmill, any size will do,
and any Dude that has th' price can buy one anywhere.
Just go to town and pick it up and haul it right on back,
then dig some holes around th' well and put it in th' air

But like a kid a-ropin' on a shetland, he can overmatch th' load,
and he better do some figgerin' so he'll kinda get th' drift,
about how far down it is to water, and th' cylinder size
to get th' water that he needs, and th' load there is to lift

These iron hearted critters come in assorted sizes,
from six through sixteen feet in increments of two,
and a tower to match th' weight of th' mill,
so it will work when he gets through

First let's talk about windmills and towers. Now I know
well and good that anyone can buy a windmill and tower, but
I'm going to give you a few hints anyway. Might even save you
some money! The first hint is, if it's at all possible, use th'
wife's money. If you've managed that, then you're off to a good
start.

The first thing you're going to have to know is just how
much water you will need. Here is a place where it's better not
to guess, and after you make a reasonably good estimate, then
you should add about ten percent. It's easy enough to use less
water than the capacity of your water system, but it's pretty
hard to pump and store more water than either the well or the
mill or the pump can deliver. I should mention at this point
that the estimate you are about to make holds up just as well
for a windmill or for an electric pump.

AVERAGE WATER NEEDS

Type	Gallons
Milking cow, per day	35
Dry cow or steer, per day	15
Horse, per day	12
Hog, per day	4
Sheep, per day	2
Chickens, per 100, per day	6
Bath tub, each filling	35
Shower, each time used	25 - 60
Lavatory, each time used	1 - 2
Flush toilet, each filling	2 - 7
Kitchen sink, per day	20
Automatic washer, each filling	30 - 50
Dishwasher	10 - 20
Water Softener	up to 150
¾-inch hose, per hour	300
Other uses, per person per day	25

— Courtesy of Aermotor Division, Valley Pump

This table — made up by the Aermotor people for use by professional windmillers — is all you need to make your water use estimate, whether the mill or pump is to be put up in some remote corner of your ranch to water a few head of cattle or for domestic use. Note that these are daily water needs (except in the case of baths and things like that where you have to figure out how many each person in your family will take).

Another thing you need to know is the amount of water your well will deliver. It won't do any good to install a mill or pump that will deliver 325 gallons an hour when your well will only give you 200!

While you are figuring out what you'll need, there are some other things to consider. One is the depth of the well. Windmills and pumps are designed to operate at certain depths, and what's really important here is not how deep it is to the bottom — although you've got to know that, too — but how far it is down to water; the static water level. If you have a two-hundred foot well but you've been able to prove that the static water level is only about a hundred feet down, then you'll save some money because you can use a smaller mill and shorter pipe and sucker rod strings. But there's still more. A lot of wells, especially here in the Southwest, just won't deliver a lot of water. A well that will give twenty gallons of water per minute is practically unheard of except in river bottoms, and a well that'll yield fifteen gallons is considered a strong one. There are a lot of places, though, where if you get a gallon a minute — year round — then you can consider yourself lucky. So, even though you have a high static water level, you may find that because the well is weak it'll draw down quickly. The water level that was at a hundred feet in your two-hundred foot well may drop fifty or even a hundred feet. If that's the case, then you'll have to install more pipe and sucker rod because you'll want your cylinder to be able to pump as much water as there is in the well.

The more that is known about the well depth, amount and size of casing, static water level, drawdown level, and rate of recovery — the better. Do not take anyone else's word for the depth and static water level. It's not that people are dishonest about these facts, but often times they are simply mistaken. Measure both the well depth and the static water level yourself — see page 30 for a quick and easy way to do this.

Bear in mind that we are only going to be lifting the weight of the water from the static water level to the top of the well, but that the cylinder must be set below the expected drawdown if possible. The cylinder is the most important part of the pump, in fact, it is the pump. Later I will go into more detail about the many different kinds of cylinders, but for now we'll just consider the all brass ball valve (ABBV) type. As I mentioned, we should try to establish the fact, by means of the bailer drawdown test, that we have enough water to pump with a windmill.

Below is another table that will help you out. After you've determined all those things I just talked about — water needs, well yield, static water level, drawdown rate, and well depth — then you can start figuring what you'll need in the way of a mill. (We'll cover pump selection in more detail a little later on.) When you're figuring the size of your mill, as a general rule you can estimate that the mill will pump the equivalent of four or five hours of full capacity out of every twenty-four, although this varies with the locality. This table is pretty much self explanatory and easy to use once you've determined just how much water you'll need every day. I should just mention here that when considering "elevation," the figure you must use is the distance from the lowest water level in the well to the discharge level above ground.

AERMOTOR PUMPING CAPACITY

Diameter of Cylinder (Inches)	Capacity per Hour, Gallons		Total Elevation in Feet — SIZE OF AERMOTOR					
	6 Ft	8-16 Ft	6 Ft	8 Ft	10 Ft	12 Ft	14 Ft	16 Ft
1¾	105	150	130	185	230	420	600	1,000
1⅞	125	180	120	175	260	390	560	920
2	130	190	95	140	215	320	460	750
2¼	180	260	77	112	170	250	360	590
2½	225	325	65	94	140	210	300	490
2¾	265	385	56	80	120	180	260	425
3	320	470	47	68	100	155	220	360
3¼		550			88	130	185	305
3½	440	640	35	50	76	115	160	265
3¾	—	730	—	—	65	98	143	230
4	570	830	27	39	58	86	125	200
4¼		940	—	—	51	76	110	180
4½	725	1,050	21	30	46	68	98	160
4¾	—	1,170	—	—	—	61	88	140
5	900	1,300	17	25	37	55	80	130
5¾	—	1,700	—	—	—	40	60	100
6	—	1,875	—	17	25	38	55	85
7	—	2,550	—	—	19	28	41	65
8	—	3,300	—	—	14	22	31	50

Capacities shown in the above table are approximate, based on the mill set on the long stroke, operating in a 15 to 20 mile-an-hour wind. The short stroke increases elevation by one-third and reduces pumping capacity one-fourth.

— Courtesy of Aermotor Division, Valley Pump

You still have to consider tower height. According to the Aermotor people you should select a tower that's high enough to place the wheel at least fifteen feet above all surrounding wind obstructions, like buildings and trees, within a radius of about 400 feet. But there are some mechanical considerations, too, that an Aermotor book doesn't mention. The tower has to be high enough so that — mechanically, for maintenance and installation of pipe and sucker rod — it will enable the people working on it to clear a twenty-one-foot length of pipe (the standard length), the elevators, and the snatch blocks. For this reason, a tower of at least thirty-three feet is best. Here's the last of the tables, again from Aermotor, showing you what sizes of towers you can buy and just what their dimensions are. All this information will come in handy, especially when you are digging the holes for the tower legs.

The tower spread is approximately one foot to every five feet in height. When anchored securely the tower will stand through storms that will wreck buildings and trees.

BASE DIMENSIONS . . . 4-POST TOWERS

STANDARD TOWER					WIDE-SPREAD TOWER											
FOR 6, 8 and 10-FT. MILLS					12-FT. MILLS				14-FT. MILLS				16-FT. MILLS			
HEIGHT	A		B		A		B		A		B		A		B	
(Ft.)	Ft.	In.	Ft.	In.	Ft.	In.	Ft.	In.	Ft.	In.	Ft.	In.	Ft.	In.	Ft.	In.
21	6	0	8	5 13/16	—	—	—	—	—	—	—	—	—	—	—	—
27	5	6 13/16	7	10 1/2	6	8 1/8	9	5 5/16	6	8 1/4	9	5 1/2	—	—	—	—
33	6	10 3/4	9	9	8	4 1/4	11	9 3/4	8	4 3/8	11	9 15/16	8	5 1/2	11	11 9/16
40	8	2 11/16	11	7 1/2	9	11 7/8	14	1 1/2	10	0	14	1 11/16	10	1	14	3 1/8
47	9	6 5/8	13	6	11	8 1/16	16	6 1/16	11	8 3/16	16	6 1/4	11	9 1/8	16	7 9/16

— **Courtesy of Aermotor Division, Valley Pump**

Before you buy a mill or pump, be sure and give some thought as to where you are going to buy it and who is going to install it. Unless you have the tools and experience, and just plain want to work, your best bet is probably your local wind-miller. A good poiŋt to keep in mind is that when you take off for parts unknown and buy a windmill, as soon as you pay for it, barring factory defects, it's your windmill. While it's true that windmills nowadays are made out of steel, they won't take a lot of abuse. Let the windmiller take that responsibility and you'll probably be better off.

You should also consider your storage requirements. My one rule is — the more the better. For any home application a six-foot diameter, eight-foot tall round tank is usually adequate. This size tank is easier to handle, and requires no road overwidth permit during transit. If one isn't enough, then ııse two. As for livestock, a steel rim thirty feet in diameter and five feet high, filled with a windmill that is always kept in the wind, will water all the cattle in the vicinity that southwestern range-land will carry. The rim should always be equipped with an adjustable standpipe, that is, one that is made of flexible pipe or one that is hinged. Either one will enable you to adjust the height of the water. For desert areas a large-diameter rim will allow more water loss through evaporation. Strive for the least diameter and highest rim possible.

Regardless of who buys the windmill, or where it's bought, here are some things to look for before loading up a new one.

In most cases windmills will not be picked up at the factory — especially Aermotor mills, which are made in Argentina these days. They'll have to be shipped somewhere, many times I suspect to the moon and back. Stevedores and freight handlers, being what they are, are not really interested in whether or not you pump water. So, if there is any sign of crate damage, especially to the crated motor, unpack it and check it right on the spot. Things to check for are a bent or damaged helmet, cracks in the motor case (check this one carefully as sometimes there'll be hairline cracks that are barely visible), a chipped turntable, and large gear alignment, where one of the gears is running ahead of the other. There are still other things to look for, such as checking to insure that the pin for the pump yoke rod is in place, that the knob on the replaceable bearing — between the two small gears — is fitted into the slot in the case, and that the hole in the case for oiling the mast pipe is open. Make sure, too, that the oil ring is neither bent nor broken and that the stroke setting on both sides of the pitmans is the same, that is that for the long stroke setting that goes to the top hole on the pitman and the outside boss on the large gear, or for the short stroke that the bottom hole on the pitman arm is used with the inside boss. (The inside boss is closer to the center of the gear.) And last but not least, make sure that all wheel-arm holes in the hub are tapped. (It's a lot easier to tap one of these on the ground than it is when you're forty feet off the ground during a good wind while hanging on to a wheel segment.)

There isn't too much to check for the unassembled tower. Just make sure all the parts are there and that the tower bolt threads and the nut threads are the same — either all cut or all rolled.

Now, lets spend your wife's money and equip that two-hundred foot well we've been talking about. As long as it's her money, here's a list of all the items you'll need, at 1980 prices.

10' Aermotor Windmill	$1,999.00
33' Aermotor Tower	1,259.50
189' of 2" reamed and drifted galvanized pipe @ 2.72	514.08
190' of ash sucker rod @ 1.61	305.90
6'x 8' 1690 gallon storage tank	750.00
1⅞ x 30" all brass ball valve cylinder	155.28
Miscellaneous fittings	35.00
1 gallon windmill oil	1.50

That's just for parts! You total it up. I don't have the heart.

IV

WINDMILL INSTALLATION

Making Water Out of Wind

The wild wind speaks among the peaks,
and sweeps across th' desert floor,
the tumble weeds fly up in th' sky
and have for evermore.
The snakes and rats in their habitats,
wait patiently for th' rain
that the winds may bring in the early spring,
to give them life again.
But the cattleman in his chosen land,
in th' rain he may not trust
if he's to survive and keep his cattle alive
then water is a must.
So he sends for a rig and a man to dig.
He'll gamble on a well.
He'll bet his soul on an eight inch hole,
here on the lid of hell.
Now the water he found down in th' ground
is of very little worth,
it must be pumped and then be dumped,
in a storage here on earth.
So a windmill stands here on his land,
and th' wind it turns th' crank.
His gamble paid and he's got it made,
as the wind fills up his tank.

There are a lot of ways to install a windmill. At this point there is just one thing I'd like to say, and you'll have to remember it all through the installation process. The mill has to be level and plumb over the well-pipe when the job is completed.

Since we are talking about level and plumb, now seems to be a good time to mention a couple of things about wells. Good well bore alignment simply means whether or not a well is either plumb or crooked. (One could possibly be straight and still not be plumb.) Although a rod pump will operate in a well that may be out of plumb, it will certainly require more maintenance, consisting mostly of replacing worn sucker rod and the water column.

Now that I've gotten that off of my mind, let's proceed with the job. The first thing you have to do is assemble the tower, and there are two ways to do it. You can construct the tower from the ground up, just like a building, over the wellhead. Or, you can assemble it lying on the ground, and pull it up — erect — over the well.

Before you begin assembling the tower — in this case we'll try it from the ground up — it's a good idea to uncrate the thing and make sure you have all the parts. First — on the ground, of course — place the posts in the proper sequence. The top of each post is the end with three holes in it. Next, place the angle iron "x" braces in a group (these braces should be installed so the outside of the angle faces up when the tower is erected); then the girts; then the wire braces, with the longer braces first because you'll need them first; then the nuts and bolts. Three or four coffee cans are good containers for the nuts and bolts. Usually, there are two different bolt lengths. Use the shorter bolts wherever possible, such as in the post splices. The longer bolts will be needed for the "x" braces. All the nuts will be the same.

When assembling the tower from the ground up, start by bolting the pads into the anchor posts. Then assemble the bottom section of the tower horizontally on the ground. Measure on the sides and diagonally between the legs to find the correct distance from the center of the well for the anchor holes. These anchor holes should be dug about four feet deep, level at the bottom, and large enough to enable you to shift the tower around to get it square and plumb over the well head. After the holes are dug, turn the bottom section of the tower upright, and place the pads and their legs in the holes. Using a couple of good 2 x 8-or 2 x 10-foot boards as a working platform on each assembled tower section, go on up with the tower until you have it all put together.

I guess you could pretend that it's a giant Erector Set — it's about that simple. I don't recommend cementing the pads into the anchor holes until the tower is level and plumb on top, like it should be. It's also not a good idea to completely tighten all the bolts until the tower is up and the mast-pipe is in place. Doing it beforehand makes it tougher to adjust the tower for

The pad is bolted to the anchor post.

proper placement over the well-head. After the bolts are tight, it is time to square, level, and plumb the tower. Then you can cement it in place.

Six- and eight-foot windmill motors can be manhandled up the tower easily enough, but for larger mills, a gin pole is needed. You can make a gin pole yourself, but it requires a welding rig and some pipe. I use a good twenty-one-foot length of four-inch pipe with collars welded at each end. A four-inch x five-eighth inch slot is cut through the collar and pipe end at the top to accommodate a piece of one-half-inch boiler plate with a hole cut in to hang a snatch block in. On top of this is a flat plate eight or ten inches square with loops on each corner for guide lines. Through the center of this plate is an eye-bolt to hold the snatch block bar in place. Perhaps you might be wondering why so much rigging and why it's so heavy? The only answer I have for that is that I have never dropped a windmill motor.

Your gin pole can be pulled up the outside of the tower with a truck by using two snatch-blocks, the top block tied to the mast pipe and the bottom block to a tower leg at ground level. Run a cable through the snatch-blocks and tie it to the bottom end of the gin pole. Then tie the cable with a rope or small chain around both the gin pole and the cable, about two-thirds of the way up the gin pole. This will keep the gin pole vertical when you're pulling it up the side of the windmill tower. Now pull this last tie up close to the snatch-block, put an adjustable chain around the mast pipe and gin pole, take the last tie loose, put on the gin pole cap and guy lines, and pull the gin pole on up to the top. When it is high enough to allow a motor to clear the mast pipe, tie chains around two tower legs, set the bottom of the gin pole in the chain hooks, tie the bottom to the opposite side of the tower so that it can't kick out, and set the guy lines. The top block and cable that was used to set the gin pole may now be put in the snatch block bar and used to pull the mill up. When tying on the mill, be sure it is balanced before it is raised. If the tail assembly is installed before you raise it, then tie the mill in the open position. If you don't do this, and the mill should happen to fold up, then the balance will be lost. When the mill is started down on the mast pipe it must go on absolutely straight or you may well bend the mast pipe. Centering the mill over the mast pipe can be accomplished by using the guy lines. Be sure to put the furl ring and arms on the mast pipe before you lower the mill on, or you'll have to do the whole thing over. If you raised the motor assembly without the tail, then lean the gin pole far enough away from the tower to balance the tail when you are installing it. After the tail is in place, take the gin pole down by reversing the procedure you used to put it up.

Now is the time to hook up the brake system. Take one furl arm loose from the furl ring and put the end with the knob on it in the socket of the tailbone casting. Put the other end back in the furl ring. The other furl arm is attached to the brake lever. The best way to install this one is to take the brake lever off of the motor, slip the brake lever socket on to the furl arm knob, then put the brake lever back on the motor. Now you can install the furl link, the rod that connects the brake lever to the tail bone casting.

Now put on the furl stick (lever) and furl wire. Adjust the wire so that when the buffer device is pulled against the motor case the wire can't be pulled any tighter. If you allow the lever

(Note — for this and all subsequent part mentionings, please refer to Diagrams on Page 47 and 48.)

43

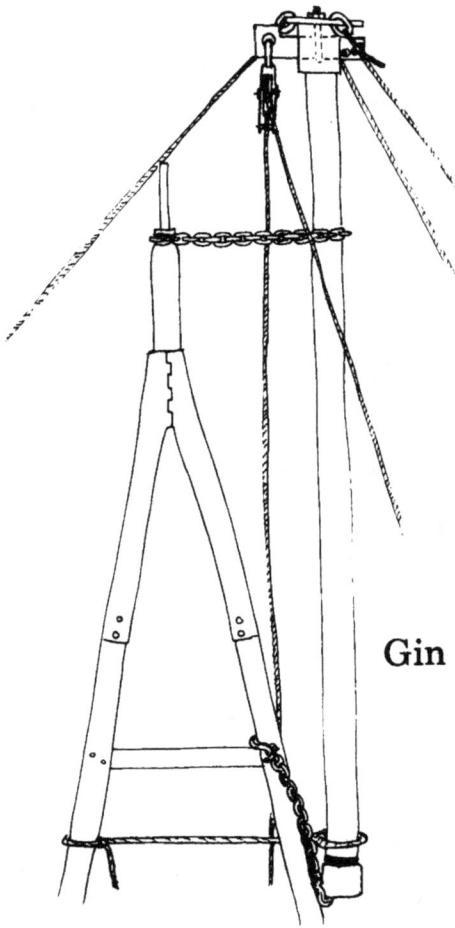

Gin Pole

and wire to pull too tight, then it's possible to bend or break the furl arms.

This may be the place to explain the advantages of the Aermotor braking system. First and foremost, all of it is outside the mast pipe. There is no pullout chain and sleeve to wear against the pump rod. A windmill with a pullout chain in the mast pipe is a potential wreck. If the chain wears in two or breaks, and jams the pump rod on the down stroke, then something has to give. What you usually end up with is a disaster.

The stick lever for the Aermotor system should be where it can be reached from the ground. It should be bolted to a tower

Slots are cut into the collar and pipe at the top.

The plate for the guide lines and the snatch block hanger.

An Aermotor stick lever is better than this winch.

post that is on the opposite side of the tower from the top lever so that the furl wire pulls across the tower. Many people use a gear and winch arrangement in place of the stick lever. This is OK if you know when to stop cranking, but young boys, bird hunters, or tourists don't usually know when this point is reached, and often keep on cranking only to break the furling mechanism.

Next, hang a hand line on the tail of the windmill, then come on down from the tower and find the wheel. You may locate it in a crate that looks like grandma's luggage after she's been on a Caribbean cruise. Nevertheless, that's the complete wheel, knocked down and laid out flat in the box. Theoretically, all you have to do is bolt 'er together.

Now is the time when you might want to take a break and rake the engineers, technicians, and the manufacturer over the coals for not assembling the wheel sections — using rivets — before they ship it. They all holler "shipping costs," and things like that, and that will be all the satisfaction you'll get. So, you might as well just get started putting the thing together.

First, slide the sail ties on sails, then slide the sails on the outside band. You'll find that the bolt holes for the sails are farther from one end of the band than from the other. Be sure and put the long end of the band to the outside of the sail. The same goes for the inside band.

After the wheel sections have been assembled, clean all the

A wheel section.

WINDMILL REPAIRS AND REPLACEMENTS

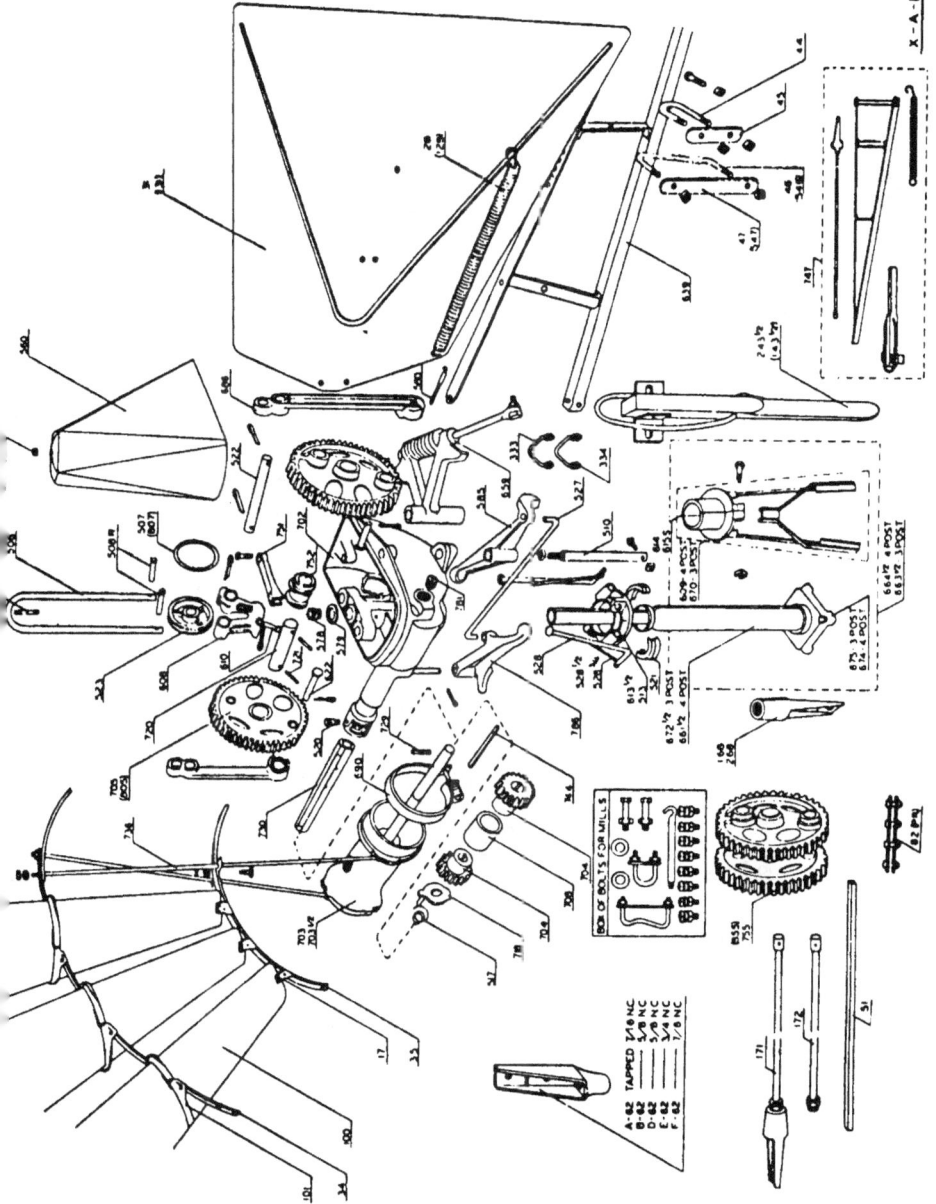

REPAIRS AND REPLACEMENTS

6-FOOT (X)	8-FOOT (A)	10-FOOT (B)	12-FOOT (D)	14-FOOT (E)	16-FOOT (F)

USE PREFIX LETTER TO INDICATE SIZE OF MILL.

PART NO.	DESCRIPTION	PART NO.	DESCRIPTION
17	Sail Tie	576	Steel Washer For Turntable
28	Vane Spring	578	Locknut For Top Of Pipe
31	Vane (Also 131)	579	Lockwasher For Top Of Pipe
34	Outer Band Of Wheel	580	Vane Spring Holder
35	Inner Band Of Wheel	582	Furl Wire — 25' Long
44	Narrow U-Bolt and Washer	585	Tailbone Casting
45	Washer For 44	588	Stud For Pitman Guide
46	Wide U-Bolt and Washer	608	Pump Rod Yoke
47	Washer For 46	609	Furl Lever Complete — 4 Post
51	Wood Pump Pole — 14' Long	610	Pin For 608 Yoke
62	Connection, Pump Pole To Well Rod	613½-O	Upper Furl Ring Without Furl Arms, S
82	Pair Splice Straps With Bolts	614	Split Furl Ring — 4 Post
100	Sail With No. 17 Riveted On	615-S	Split Furl Ring — 3 Post
100½KD	Section of 3 Sails	622	Bolt With Cotter
	3 No. 100, 3 No. 101, 1 No. 34	639	Tailbone With 510 Pivot
	1 No. 35, 10 Bolts, 10 Locknuts	659	Buffer Device — Complete
101	Sail Rib	661½	Pipe With Base — 4 Post
101N	Locknut for 101	663½	Pipe With Base, Furl Lever — 3 Pos
128	Vane Spring	664½	Pipe With Base, Furl Lever — 4 Pos
131	Vane	670	Furl Lever Complete — 3 Post
143½	Furl Handle	672½	Pipe With Base — 3 Post
168	Swivel Casting For Pump Rod	674	Pipe Base Only — 4 Post
171	Pump Rod w/Swivel Casting (complete)	675	Pipe Base Only — 3 Post
172	Pump Rod w/Swivel Nut Only	686	Pitman Arm — Adjustable Stroke
243½	Furl Handle	690	Brake — Complete
268	Swivel Casting For Pump Rod	702-O	Main Frame
333	V-Bolt For Connecting Furl Handle	703-O	Hub and Shaft
	Tower Corner Post	703½-O	Hub and Shaft
334	V-Bolt For Connecting Furl Handle Tower	704	Small Gear
507	Oil Ring	705	Large Gear
508	Pitman Guide With 588 Stud	708	Replaceable Babbitt Bearing
510	Pivot Bolt For Tailbone	718	Spout Washer
513PP	Split Upper Furl Ring	720	Shaft For 705 or 805
517	Spring For Spout Washer	721	Pin For Large Gear
520	Oil Collector In Hub	729	Pin For Shaft On Hub
521	Split Washer	730	Replaceable Sleeve Bearing
522	Shaft For Guide Wheel And Yoke	736	Wheel Arm
523	Guide Wheel (for Pitman and Yoke)	744	Key
527	Furl Link	747	Tailbone Bundle
528	Furl Arm	751	Bearing Bar
528½	Furl Arm, Long For Repair	752	Bearing For Large Gears
	(Takes Up Wear On Brake)	755	Large Gear and Bearing Assembly
528¾	Furl Arm, Extra Long For Repairs		2 — 705 Gears, 1 — 720 Shaft
	(Takes Up Wear On Brake)		1 — 752 Bearing, 2 — 721 Pin
546	Wide U-Bolt and Washer	781	Countersunk Plug
547	Washer for 546	786	Brake Lever
560	Helmet	799	Sails, Ribs, Wheel Arms, Bands and
565	Nut For 588		Bolts for Complete Wheel, and
			Furl Wire

wheel arm threads good with a wire brush. The arm ends with the straight and shorter threads go in the hub.

Oil the threads on the hub, and start screwing in the arms. Take extreme care not to twist the threads off in the hub. Believe me, this can happen before the arm "shoulders up," and the cross ties are in line. If the threads begin to get tight, yet there are quite a few left, back the screw out, clean the threads carefully, and try again. In lining up the cross ties on the arms, instead of bearing down hard on the wrench, try using less pressure and simply bump the wrench handle instead. If this fails, back off one thread or so or try another hole. Only the Lord knows, and He won't tell, just how much pressure on these threads is too much. But twisting them off in the hub creates a situation that is definitely not good.

Now that you have everything assembled, pull the wheel arms up the tower with the hand line. After they are installed, you're ready for the wheel sections.

Before installing the sections, turn the mill to the top of the stroke. This will be of some help turning the wheel to put the sections in. On larger mills it's a good idea to put a small come-along around the hub to help lift the individual wheel sections into place after you have taken them off of the hand line. When putting in the sections always be sure that the wheel arms are crossed. If all the arms are not crossed, the wheel will be out of balance.

Install the first section, then turn the hub till that section is dead center at the top. Being careful not to bend the wheel arms, put in two more sections at the bottom. The weight of the one at the top will help pull those that were just installed upwards. Now go ahead and put in the rest. The last section will need to go on ahead of the first one you put in so that the laps will all be the same. Do not tighten the arm nuts until all the sections are in place. Sometimes the last section can be a bear-cat to put in: usually the wheel will need spreading. If so, turn it to the top — this is where the banana and coconut diet comes in handy — and get up there with it to do the job.

After the sections are all in place, run the wheel arm nuts up to within two or three threads of being tight. Then take a spud wrench or a drift punch and a couple of pairs of vise-grips, line up the holes in the wheel arm cross ties and the inside band, clamp them with the vise-grips, and put in the bolts.

After these bolts are in place, you can finish tightening all the other bolts and nuts.

The pump rod is next. Pull the small pin out of the pump rod yoke and clean the paint off the pump rod knob. (It may even need dressing down with a file.) It's also a good idea to bevel the end of the pump rod pin to some extent. Shove the pump rod up through the mast pipe and motor case and in to the socket in the base of the pump rod yoke. (You might consider holding the pump rod in place with your vise-grips resting on the motor case while turning the gears so that the pump rod yoke comes down over the pump rod knob.) Once the two are in place, put in the pump rod pin and secure it with a cotter key.

Now for the oil. There is an oil level mark on the outside of the case marked *acyte*. (Remember, these mills are now made in Argentina and *acyte* means "oil.") Fill to this level with ten weight non-detergent oil. Do not use motor oil. The detergent in motor oil may leave a residue that will prevent the oil from returning from the oil collector to the case. As long as the big gears are running in oil, the mill will keep itself lubricated.

When you have the helmet on, you have just about finished assembling the mill. Check the helmet first to make sure there are no dents in it that the pitmans can rub against. Also check to make sure that the groove on the helmet bottom fits snug against the flange of the case. If you want to shoot a hole

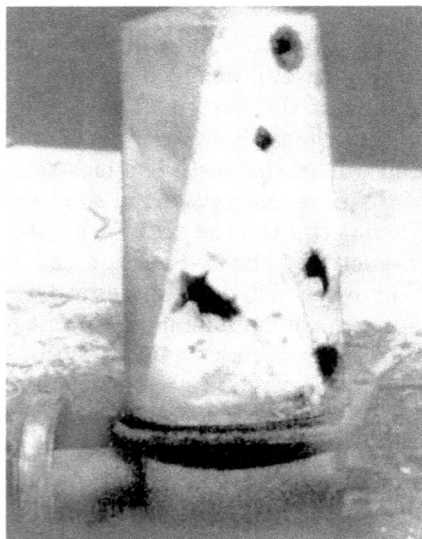

Helmets are often damaged by so-called sportsmen.

50

in the helmet before the sportsmen do — and I use that word loosely — then that's up to you.

If you are sure that the concrete around the tower legs has set long enough, your windmill is now ready to be unfurled, or in range language, to "turn 'er loose and watch 'er gin," even though the cylinder and sucker rod have not yet been installed. As insignificant as this operation may seem, there is a right way and a wrong way to do it. The mill should be eased into the wind gradually, at least until some of the tension is out of the tail spring. If you just turn it wild loose, it's possible to pop the tail off, or break the case if the wind happens to be just right. Remember — turn it on slowly.

Now that you have the mill up and assembled using the "from the ground up" method, let's try another way by assembling it completely along the ground. Anchor holes should be dug as before. If the tower is to be raised with a gin pole and a truck, it should be assembled with the bottom square with the anchor holes. Leave enough room to pull the tower up from the opposite side with a truck.

Start by putting only one side together, with the exception of the anchor posts, from the bottom to the top. It's easier if you put this side together, complete with mast pipe and with the outside facing up, then turn it over and build the rest of the tower on top of it. If the tower is to be raised with the mill in place, block the top end of the tower up off the ground and put the mill on. The bottom legs will also need to be blocked or hinged so that they cannot slip. These legs may have to be stiffened with pipe, depending on the height of the tower and weight of the mill. When tying the pull cable to the tower, do not tie it to the top of the tower. It's better to make the tie about three-fourths of the way up, so there is less strain on the legs. A gin pole and snatch block for the pull cable can be put at about the center of the tower to get it started up. Also tie on a couple of guy lines to guide it and keep it from going too far after it's all the way up. When the tower is getting close to the vertical, or as soon as there is room to put the anchor posts in the holes opposite the hinge or the blocks, bolt them on and set the tower down on them. Then take the hinge or blocks off of the other side and bolt the anchor posts on to that side. Now you can square, plumb, and level the tower, then cement it in.

For raising the tower and mill with a mast rig such as the

Hydra-hoist, the tower and mill should be assembled in the same manner as before except that it must be laid at a right angle to the mast rig. The rig has to be centered square with two of the anchor holes. When tying on the tower, make sure there is enough room between the tie and the mast sheave for the tower to clear the ground. When the tower has cleared the ground, turn it one quarter of the way around toward the mast rig and bolt on the anchor posts. Using this method there is no strain on the tower legs, and it is the safest and easiest way I know to put up a windmill—if you happen to have a mast rig.

Now that we've done such a good job of getting the windmill put up, we are ready for the pump. As insignifcant as installing the pump might seem, this is a time when a lot of mistakes can occur.

One of the first things we should consider is where we are going with the water. We have to consider whether the water is to go to an overhead storage tank, a steel rim, an earth tank, or over a hill to another location. It is important to know ahead of time whether the lead-off pipe is going to be above ground or whether it will be buried. Then, too, we have to take into account any problems that your location might present.

Since you've already tested your well to make sure of its static water level, its depth, its drawdown level and rate of recovery, you can go ahead and install the pump, confident that the well has enough water in it to get the job done.

Bear in mind that we are only going to be lifting the weight of the water from the static water level, but that the cylinder must be set below the expected drawdown level, if possible. The cylinder is the most important part of the pump —in fact it is the pump. A little later on I will go into more detail about the various kinds of cylinders, but for now we will use the all brass ball valve type. Because we know from testing that we have more than enough water to pump with a windmill —that the drawdown level is well over the top of the intended location of the cylinder—we don't have to consider installing a tail pipe or strainer on the bottom of the cylinder.

For the water column we are going to use two-inch galvanized reamed and drifted pipe. Although reamed and drifted pipe costs a few cents more per foot than standard pipe, it offers several advantages. It is straighter and cleaner on the inside, thus it accumulates less rust and corrosion. Most impor-

52

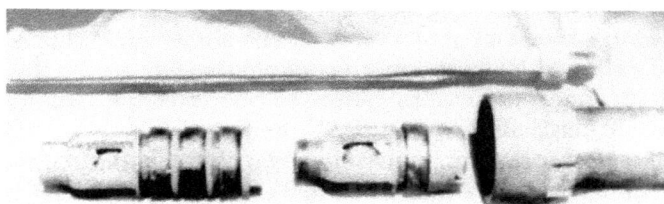

The item on top is the sucker rod connection.

tant, the longer collars afford more thread protection and holds much more weight because you can use more thread. Since the collars on new pipe were only tightened at the factory by hand, remember to tighten them before you put them into the well. If you don't, you could wind up fishing for the whole mess.

Galvanized pipe for the water column is generally considered to last longer than black pipe, however, in some cases this is not true. I do not really know why. I just know. I suppose that an analysis of the water might determine which would be the best one to use, but mostly I have gone along with the trial and error method. If I'm not satisfied with the life of galvanized pipe, then I try black, and in some cases black will outlast galvanized three to one. In any case, I've seen water column pipe corrode in a couple of years or so, so be prepared to have to change it at least every once in a while. On the other hand I've also seen it last thirty years. I might add, too, that my experience with cheap foreign pipe has made me appreciate the "American Way."

The wood sucker rod that we are going to use will be ash. As far as I know, this is the best that can be bought. There are other kinds of wood, but I don't recommend using them. One big advantage of a wooden sucker rod is that it will float, and it is lighter than metal rod. Also, there is a lot less rust. Rod connections are galvanized and fastened to the rod with copper rivets. Incidentally, nails or bolts should never be used to replace the rivets because they will rust out and drop off into the cylinder. A word of caution here. Pick a sucker rod that has good straight connections for the one that you intend to use at the bottom, attached to the plunger valve. If you don't, and the connection is off center, then the leathers on the valve will wear unevenly. The largest wood sucker rod that should ever be used in a two-inch pipe is one and one-eighth inch; this is the size we will use. The top rod, the one that is exposed to the air, should be made of metal to eliminate wood rot that usually affects that particular section if it's made of wood.

The first operation in installing the pump is to run (install) the cylinder and pipe (water column). If the Hydra-hoist is to be used we can be ready in about twenty minutes. Simply back up to the tower, raise the mast high enough to clear a length of pipe plus the pipe elevator, drop the swivel hook down through the tower, and we're ready.

If you use snatch blocks, cable, chains, and a truck, you will have a little more getting ready to do. Tie a good chain near the top inside of the tower across two legs on one side, and hang a snatch block on it. Never do this diagonally across two legs. Many towers have been bent and therefore weakened by tying the chain diagonally across two opposite legs. True, when the block is hung on our arrangement, it will not be exactly centered over the well, but it's better than a ruined tower. Now we must have another snatch block near the ground. This block may be tied to the casing or to a tower leg on the opposite side of the truck. A three-eighth inch single cable with a swivel hook and safety latch, long enough to thread through the two blocks and tie to the front of the truck, will easily handle the weight of this particular installation. For the heavier loads of bigger windmills, a longer cable may be used by tying the end to the top of the tower and using a third block for a travelling block, thus doubling the line and increasing the weight carrying capacity.

Now is the time when the handyman or the do-it-yourselfer may prefer to use a set of two by two rope blocks. If the entire family and some of the neighbors are available to handle the weight of the last few lengths of pipe as you lower the whole pipe string into the well, that's OK. These blocks should most certainly be strung with enough rope to handle a full length of pipe, and that's a lot of rope in a country where pack rats are common.

The standard tower does not come equipped with a built-in A-frame to allow for getting pipe and rod inside the tower. The next best thing is to take the first girt and X-braces loose on the side where the pipe is laid. Now we are ready to run the

The O'Bannon cylinder.

Rope blocks.

pipe and the cylinder. Set the cylinder in an elevator under the collar of the first length of pipe, always with the latch up. Then hoist away. When the pipe is vertical in the tower, clean the threads thoroughly with a wire brush and dope them good with joint compound, use roofing tar mixed with graphite — anything is better than nothing. After we have screwed the first joint into the cylinder and lowered the first length into the well, we just continue adding pipe joints until we have the cylinder at the proper depth.

Many times a few more feet may be desired for the depth of the water column. This, of course, may be accomplished by cutting and threading the water column and sucker rod. However, a less expensive and easier way is to simply add a tail pipe to the bottom of the cylinder.

Since I just mentioned the word "cylinder," now seems like a good time to discuss some of the better known kinds.

It is my opinion that the best water well cylinder — especially for deep wells — is the O'Bannon. Being the best is naturally being the most expensive. The initial cost usually will be about three times more than a conventional cylinder, but in most cases the advantages will more than off set the cost difference. The pumping action of the O'Bannon cylinder differs from the conventional cylinder in two ways. First, there are no cup leathers, and second, the cylinder discharges on the down stroke. These two factors lighten the load to the extent that it is generally assumed that a size smaller windmill may be used than need be used with a conventional cylinder. The pumping action stems from one or more twelve-inch plungers of Cro-lay steel working in the barrel of the cylinder, the clearance between the plunger and the barrel being about that of a human hair. Each plunger is good for four hundred feet of lift.

There are two types of O'Bannon cylinders, the rod type and the tubing type. The rod type is a complete pump, and may be installed as one unit whereas the tubing type is installed in two parts. The rod type may be installed without pulling the water column by removing the conventional cylinder valves and seating the O'Bannon in the conventional cylinder. Make the water column seal with B-13 seating cups on the bottom of the O'Bannon. If the water column is out of the well, an O'Bannon seating nipple may be used instead of the conventional cylinder to make the seal. The rod type comes equipped with seating cups for the O'Bannon seating nipple: if it is to be seated in a conventional cylinder, have the dealer replace these cups with B-13 seating cups. In the case of the rod type cylinder the barrel works on the stationary plunger held in place by the seating cups and a connecting rod that extends from the seating assembly to the inside of the bottom of the cylinder barrel. This rod serves as a guide for the cylinder barrel; the lock nut on the end of the rod is to hold the seating assembly to the cylinder when lowering or raising the cylinder in the water column.

The tubing type of O'Bannon cylinder differs from the rod type in the following manner. The barrel and the bottom valve screws on to the water column replacing the seating nipple or conventional cylinder. The plunger and top valve are lowered into the barrel by the sucker rod. The bottom of the plunger assembly has a male thread that fits the bottom valve and seal in case it has to be pulled.

If for any reason either of these cylinders are pulled out of a well in cold weather — and this is important — do not let it

56

freeze! The only way to get all the water out of the plunger core is to completely remove the plunger and clean it. If the plunger is allowed to freeze it will swell, crack, and lock. Usually one plunger can be removed and replaced with a new one, but if two or more are frozen the barrel will probably be ruined before they can be removed. This is a case where cold storage won't help at all.

As for conventional cylinders, I believe that—like washing machines—the fewer working parts the better. One of the first things that comes to mind is the so-called easy-pull deep-well cylinder. The theory and design on a desk top looks pretty good, but may be something else again in the bottom of the well.

The "easy-pull" conglomeration consists of a connecting rod similar to the rod type O'Bannon, but of a much smaller diameter, and usually made of brass. This connects the stationary valve to the plunger valve. The theory is that when the plunger valve is pulled the stationary valve has to come out too. This is true, but only if the connecting rod has not worn to the breaking point. If it has, and it breaks, the water column will have to be pulled to get the stationary valve. Another disadvantage, though not often encountered, is that if the connecting rod holds and the stationary valve will not come loose, there is no way to unscrew and pull the sucker rod before pulling the water column. I will let you use your imagination as to what might happen if the connecting rod should break during the pumping operation.

The cased cylinder is a piece of pipe the same size as the water column that has a brass sleeve for a lining. The weight of these cylinders may lead to the belief that they will last longer than the all brass barrels. Except for use with extreme pressure, such as is found in real deep wells or when forcing water through a stuffing box, this is not true. It has been my experience that the sleeves in these barrels are not as thick as those of the all brass, and several times they have been known to slip in the barrel. So for my part, I'll take the all brass barrel in preference to the cased brass-lined barrel.

Regarding the length of the cylinder barrels, naturally the barrel has to be long enough to accommodate the stroke of the windmill or pump. I consider it good policy to buy a cylinder whose length is equal to two pump strokes. What you can do is

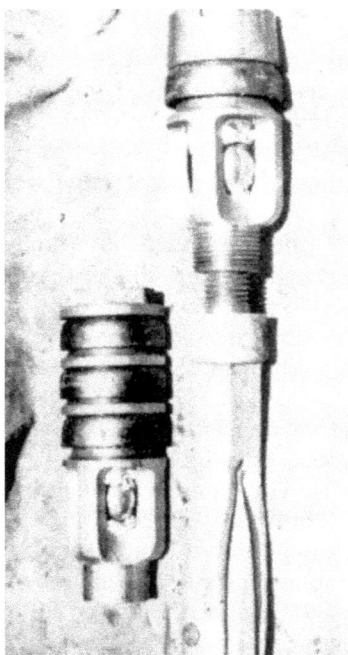

Another use for boiler plate.

Ball type valves.

set the stroke of the plunger valve in either the bottom or top part of the barrel, and when this is worn move the stroke to the other part of the barrel.

The two most common types of cylinder valves are the ball type and the spool type. Personally, I prefer the ball type as it is not as easily fouled. If it does become fouled, many times it will clean itself and go back to working. When the spool valve becomes fouled, it will usually have to be pulled and cleaned by hand. Two more disadvantages of the spool are that when the neoprene or fiber washer on the bottom of the spool becomes worn, it may leak, and when the sides of the spool and inside of the cage wear, the spool may cock.

Among some of the trials and tribulations that I have encountered in life are half-soured beans and spring loaded cylinder valves, both having about the same effect. The springs on top of the spools of these valves usually wear and break or corrode and then will not work. Fortunately, these valves are in the minority.

The cowboy's drinking fountain.

Now back to our well. We will set the last pipe on a piece of half-inch — or thicker — boiler plate with a hole cut to fit snug under the collar. This plate has to be large enough to serve as the casing seal and also hold the weight of the water column. Remember to put this plate on the pipe before the last joint is made up. From the last collar we will go on up with a nipple to where we want the tee. Many times I have seen the water column hung right on the tee. This is very poor practice for the simple reason that there are fewer threads in the tee than in the collar, thus it is weaker. If the tee and lead-off pipe are desired next to the well head, put in a close nipple, then attach the tee.

It has been my experience that the well head is the proper place for the tee and lead-off pipe. (Except when you wish to install a cowboy drinking fountain.) Let's say that we want to pump the water into a storage tank the top of which is ten feet above the tee. Many people would put a standpipe on the last collar, and put the tee on the standpipe ten feet above the well head. The disadvantage of this arrangement is that when a cowboy is working cattle around the windmill the lead-off pipe is always in his way. Also, when you need to pull the sucker rod, you have to take off the lead-off pipe and take the standpipe out. A better way is to use the close nipple and tee like I mentioned. Then run the lead-off pipe underground to the tank, as shown in the diagram following. Don't forget to put a

59

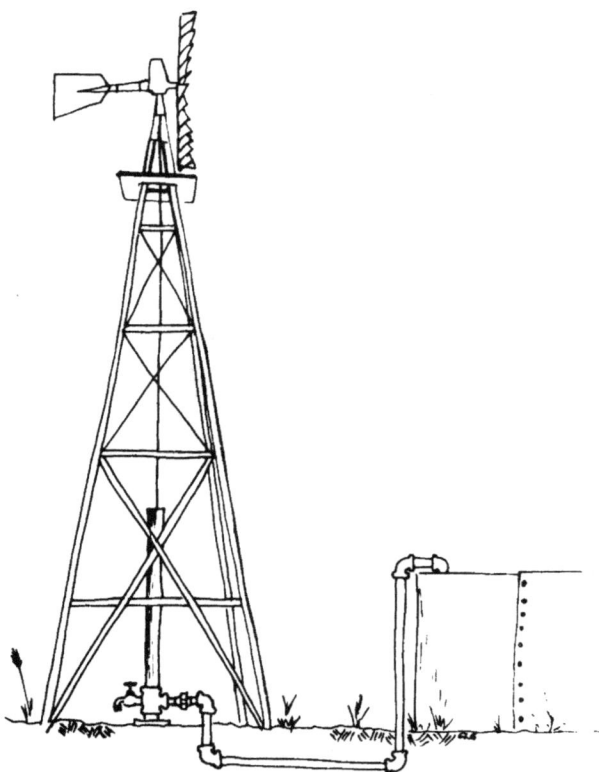

nipple and union next to the tee. You will still have a twelve or thirteen foot standpipe, but it won't be so hard to work with. You can put the sucker rod joint a foot or so above the tee, and when you get ready to pull the sucker rod you don't have to remove the lead-off pipe. Simply loosen the standpipe and pull it up so that the sucker rod joint is exposed. Put a vise or vise-grips on the sucker rod to keep it from slipping down into the well. Loosen the sucker rod joint and simply swing the standpipe and top rod to one side of the tower. Then you can work on the remaining sucker rod with no interference.

If the tank has to be set at a higher elevation than we have room for a standpipe in the tower (the standpipe always has to be higher than the highest water level in the rim or tank) a stuffing box will have to be used in place of the standpipe. But stuffing boxes are trouble. If at all possible in the planning of this operation, if it's feasible to eliminate the stuffing box, by

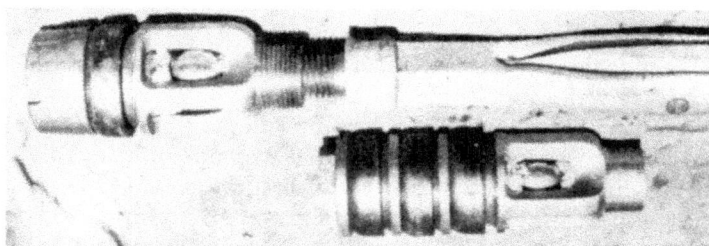

The sucker rod screwed into the bottom check valve, assuming that the plunger valve (above) has been pulled.

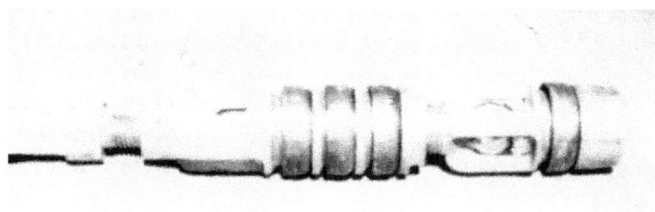

The bottom check valve is pulled after screwing it into the top one.

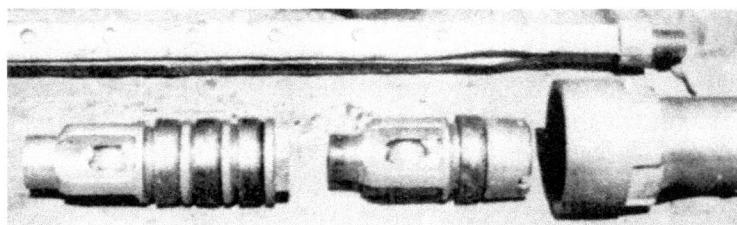

From top to bottom, left to right: sucker rod end; plunger valve; bottom or lower check valve; and cylinder.

The stuffing box.

all means do so. You will automatically eliminate a lot of headaches as well.

With all that out of the way, now is the time to run (install) the rod and the valves. We have used a one and seven-eighth inch cylinder on a two-inch water column so that we can run the valves to the cylinder after the water column has been installed. If we used a larger cylinder, the upper and lower check valves will have to be installed first and the water column and the sucker rod run together. In this method the length of the sucker rod joints and the water column joints have to correspond, and this takes some doing. The only advantage to this kind of installation is to pump more water because of the bigger cylinder. Remember, though, that when the plunger valve needs releathering, everything has to be pulled. This method of installation, although sometimes used, is most certainly not recommended.

As I have said before when I explained how a windmill works, the bottom valve is often called the bottom or lower check, and it is the stationary valve. Its one leather makes the outside seal for the water column in the bottom of the cylinder. The top of the bottom part of this valve is grooved for the ball to make a watertight seal, and has a male thread that screws into the cage. A ball rests in this cage, and when seated, it makes the inside seal for the water column. In most cases the top of the cage has two sets of threads that have nothing whatsoever to do with the pumping operation: they are used only for removing the bottom valve. The male thread on the bottom valve cage-top matches the female thread on the bottom of the plunger valve. The female threads on the inside of the male thread are the same as the female threads in the plunger-valve cage-top. The reason for all these threads is to let you pull the lower check by screwing the bottom of the upper check into it or by screwing the male sucker rod end into it (after the plunger valve has been removed), and then pulling it up. Why two ways to pull the lower valve? For one thing, it's possible that the male threads in that valve could become damaged in some way, and the extra set of threads provides an alternate way to hook onto the lower valve.

I remember one incident when because of rusted threads we failed to pick up the bottom valve with the plunger valve. My amigo from south of the Rio Grande—Señor Jose Marteen

— was with me at the time, and watched as we removed the plunger valve and screwed into the female thread of the bottom valve with the sucker rod end. When it all came out of the water column Jose pushed his straw sombrero back and muttered "Caramba. The threads, she ees got brothers, thees I do not know!"

Before running the valves and sucker rod, tighten the bottom of the valves, being careful not to use enough force to spread the leathers until they will not go in the cylinder. If it is desired to have the female end of the rod down — and some people insist that a professional windmiller install the sucker rod this way — then a male to male adapter must be used from the sucker rod end to the plunger valve.

Now comes the time to install the bottom valve. This is usually done by simply dropping it down the water column, right side up. So far in my career I have managed to avoid dropping one down a well upside down, but it has been done. And, if anyone happens to be standing around who is not familiar with what is going on, when the valve is dropped you can just say "Oh hell, I dropped it!" and this will usually get some kind of a reaction. When the valve is dropped you can hear it clanking down the water column. The first KER-WHOP you hear is when the valve hits the water. This naturally slows it down some, but by putting an ear to the tee you can hear the ball rattling as the valve travels through the water. Then comes the KER-LICK when the valve finally seats in the cylinder.

Next is supposed to come the plunger valve and sucker rod, but before I discuss their installation I should tell you a bit about sucker rods in general.

There are several different types of sucker rod. Wood rod (ash and apatong) will to a certain extent float and will therefore lighten the load on the windmill. Apatong, while somewhat cheaper than ash, should never be used in a domestic well as it will discolor the water for a week or so: if you have never seen a woman hang up a white bed sheet that has turned red, then you may never have seen an angry lady. (Moral — use ash.)

I think that it would be safe to say that the most important and most expensive parts of the sucker rod are the rod-ends, or connections. The threads of the connections are bolt threads, male on one end and female on the other. Bolt threads are fewer, but they are cut much deeper and therefore are stronger than pipe threads. The exact joint of the connections will be round, but about two inches from the joint the connection will become square. These squares are standard and will correspond to the pin size — usually the larger rod will have the larger size pin and square. It is preferable to have rod elevators for both the larger size squares and bushings and for the smaller size. This elevator will also serve as a back-up wrench for the bottom square. Any standard sixteen- or eighteen-inch wrench will do for the top square. It is not necessary or even advisable to use an extra large wrench when tightening or loosening the sucker rod joints. When the joint is shouldered. a jerk or snap on the wrench is usually sufficient, and exerting more pressure may break the pin two or three threads inside the female, and it may not be noticed. In loosening these joints it is well to use what I call a Martio-massage (sharp, light raps with a hammer) on the joint itself before attempting to loosen it.

In the case of wood rod, the connections are longer and have a yoke that is fastened to the rod with copper rivets: connections on metal rod are welded. A word of caution about the rivets. Most wood rod will come equipped with copper rivets as they are more resistant to rust and corrosion. The practice of using bolts, nails, or whatever is handy to replace these rivets should be avoided. When pulling wood rod out of a well, close attention should be given to the rivets to see that the head has not worn off and loosened the rivet. A rivet in the cylinder will usually tear up the cup-leathers, score the cylinder wall, and foul the valves. One more warning: when pulling wood rod out of the water column, if it is to be left out long enough to dry, some provision should be made to keep it from warping, otherwise it may crawl off with the snakes. Wood rod is ordered in these sizes: one-and-one-eighth; one-and-three-eighths; one-and-five-eighths; and one-and-seven-eighth inches. One of the most common mistakes made with this rod is using a size so large that it rubs the water column. Nothing larger than one-and-one-eighth inches — the smallest — should be used in a two-inch water column.

The most popular sizes for Air-Tite (metal) rod are number one and number two. Number one is three-eighth inch pipe with welded connections having five-eighth inch connections. The cost of the Air-Tite rod compared to that of plain pipe certainly makes pipe more attractive, except for the connections.

"Finger-Gitter"

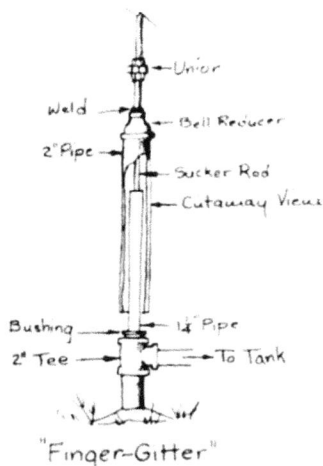

A "finger-gitter."

If pipe is to be used for sucker rod, the connections may be bought separately and welded to the pipe. Pipe collars in the sucker rod string are responsible for many fishing or pipe-pulling jobs, and therefore should be avoided.

Solid rod, the strongest and heaviest, may be bought in either twenty or twenty-five foot lengths, and will usually require a rod connection adapter for the plunger valve. In regard to the twenty-five foot lengths, consideration should be given to overhead clearance, in other words, the tower. The A-frame or whatever should be high enough to clear this twenty-five foot length.

Now let's run the plunger valve and sucker rod. Take care to select a good straight rod to screw into the plunger valve so that it will run straight in the cylinder and not cause one-sided wear on the cup-leathers. Run the rod and plunger down until the latter touches the bottom check, then raise it three or four inches. This will be the bottom of the pumping stroke.

The uppermost sucker rod will be metal instead of wood to prevent wood rot where the rod is exposed to the air. We will need to do some welding on this rod. The metal rod will have to extend far enough above the tee to allow for a finger-gitter. The finger-gitter consists of a two and one-fourth inch bushing and a one and-one-fourth by sixteen inch nipple in the tee. Over

How to make a guide.

this slip a two by sixteen inch nipple with a two by one inch bell-reducer. Leave enough room at the bottom of the nipple to allow for picking up the bottom valve when it eventually becomes necessary, then weld the bell-reducer around the sucker rod. (The professional windmiller can probably charge five dollars more for the finger-gitter if he calls it a water column protector instead!) Three or four inches above the bell-reducer, cut and thread the sucker rod an put in a union for joining the sucker rod to the red rod, often called the pump rod. (This rod is usually sent with the windmill tower, and is that part of the pumping string between the windmill and the top of the sucker rod.) The red rod is connected to both the sucker rod and the windmill pump-rod swivel. This hookup must be made with the windmill at the bottom of the down stroke. Now we need to install a guide for the red rod that is located about half way up the tower. This guide will serve the dual purposes of keeping the red rod straight and insuring that the windmill rotates about the pump rod swivel without unscrewing the sucker rod joints. A two by six inch plank bolted across two girts with a notch cut out for the red rod and rub blocks makes a quick and suitable guide.

66

The Midland and Clayton-Mark boxes.

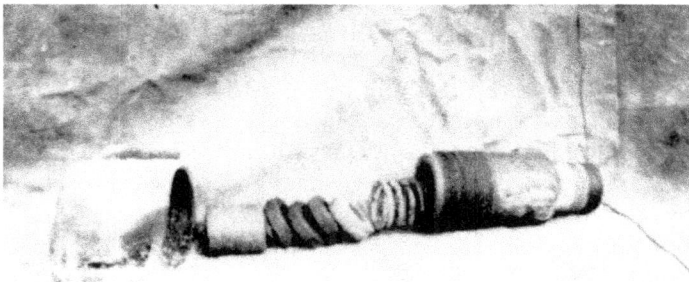

The Norris stuffing box.

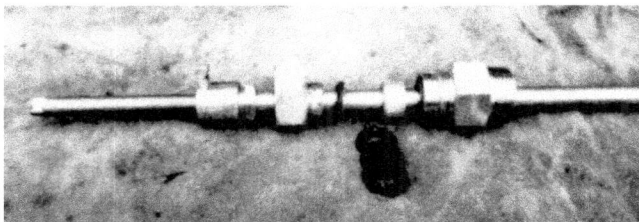

The Jensen stuffing box.

Before we end this chapter on windmill installation, I should touch briefly on stuffing boxes, packing boxes, and inverted cylinders. Many times it becomes necessary to force water to an elevation higher than what a standpipe will allow for gravity flow. In this case, a stuffing box of some kind is required. If the lift is such that it creates fifteen or more pounds of pressure at the well head, then I would suggest installing an air chamber in the service line. A note of warning at this point may be in order: do not let these boxes run dry, as this may burn the gaskets.

There are several kinds of stuffing boxes on the market, and I will list those that I have found to be the most common. These are the Norris, the Jensen, the Midland, the Clayton-Mark and the Inverted Cylinder. The Norris and the Jensen are the most popular of the brands.

The bottom of the Norris box is threaded with either one-and-one-half or two-inch pipe threads for the water column. The inside of the box consists, from the bottom up, of a spiral spring, bottom babbit bushing, four split rubber gaskets (packing rings), the top bushing, a screw-on cap, and a polish-rod from four to eight feet long. Tightening the threads in the cap presses the top bushing against the gaskets and bottom bushing, thus forming the seal around the polish-rod. This seal should be made with the water column full of water and the windmill running. When the gaskets become worn and begin to leak, they may be replaced without taking the polish-rod loose from the windmill by unscrewing the cap and raising the top bushing. When installing new gaskets, be sure and alternate the splits. When and if the bushings become worn they may also be replaced.

The principle of the Jensen stuffing box is much the same as the Norris except for the gaskets, cap, and polish-rod. The gaskets for the Jensen box are split fiber cups and they are much smaller than the Norris gaskets. The tension in this box is dependent entirely on the spring, and cannot be adjusted by the threads in the cap. The Jensen polish-rod has an adjustable brass sleeve that runs in the gaskets.

The Midland and Clayton-Mark boxes are made of brass, and the polish-rod is usually made of stainless steel. These boxes are packed with graphite rope packing.

The inverted cylinder is a brass barrel with a bushing at

the top to fit the water column. The polish-rod is equipped with cup-leathers that fit the barrel to make the seal.

At this point, now that we've covered stuffing boxes, the windmill is really ready to go. With a little luck and a little wind, we're ready to pump water, and I'd bet more on that than I would a horse race.

Like I said earlier, if there is some wind blowing, it's a good idea to ease off the brake and put the windmill into the wind slowly to make sure everything is working. Once you have determined that everything is OK, including the upper and lower check valves, then you can head it into the wind straight on. Let it pump water for a few minutes to clean out the well and pipe string. Once it starts pumping regularly, and everything is clean, then you can leave it, assured that it will do the job.

V

PUMP INSTALLATION

The best and most economical private water system that I have found is a pressure system in conjunction with either a submersible or a jet pump. The submersible is the leader of the two. It is no secret that domestic water systems are highly advertised — they can be purchased from many sources. I have seen some advertising that reads like a commercial for washing powder or breakfast food, and it may well lead you to the belief that a pump is just a pump and that's all there is to it. But it's just not true. There are several factors to be considered when you are considering installing a pump.

First and most important is the yield of the well. It is wrong to use a pump that will pump more water than the well will yield. This is probably the most common cause of submersible pump failure. Submersible pumps cannot be allowed to run dry! It is important to select a pump according to the yield of the well and the supply of water that is needed. This may be accomplished by referring to the manufacturer's performance chart. When you buy a pump, check the shipping carton and the pump for apparent damage such as dents, damaged motor lead wires, and wet spots inside the carton that might be caused from loss of lubricant.

A pump installer should have at least some knowledge of electricity. His most important tool is a combination volt and ammeter. It is absolutely necessary that the line voltage correspond with motor voltage, and you have to be able to measure it to find out for sure. The line voltage is usually AC single phase and either 115 or 230 volts, but there are often excep-

A device for lowering the pump.

tions such as a higher voltage or three-phase operation. In any case, make sure that the line voltage corresponds to the rated voltage of the motor.

Surface motors such as power centrifugal and jet pumps are usually marked dual voltage — 115v or 230v — or simply high-low. This may be misleading to some people because the terminal wires inside the motor case may have to be changed to correspond with the line voltage. There should be a diagram for the wiring either stamped on the outside of the motor case or on the inside of the terminal plate. The terminal wiring should always be checked to make sure that it agrees with the diagram.

Submersible pump motors are NOT dual voltage, and as I have said, they are usually either AC 115v or 230v single phase. All the necessary data for the pump installer, such as horsepower, voltage, and phase are printed on the motor case, and it should most certainly be adhered to.

One mistake that is common on submersible pump installations — especially for 115v motors — is using submersible electrical wire that is too small. Contrary to the belief of some, 115v motors require larger diameter wire than 230v motors. Pump manufacturers put out a chart for wire size depending on the horsepower, voltage and depth setting of the pump. If the pump installer does not know what size wire is required, then he must refer to the chart.

Penny-pinching on the cost of submersible wire is poor policy. Always use the best, and take care when setting the pump not to damage the insulation. Insulation is damaged — in most cases — due to carelessness such as allowing the wire to kink or drag over the edge of the casing. When the pump is installed in the well, the wire should be kept as straight as possible when it is lowered down the well, and it should be tied to the water column at least every ten feet. These ties should be made with waterproof tape or short lengths of submersible wire; I prefer the wire. Splices may be made in submersible wire, but they must be made water-tight. This can be done with a good grade of waterproof electrical tape, and don't spare the tape when you are doing it. I do not believe that I can stress too much the importance of protecting your submersible wire, as it is the lifeline of the pump. And speaking of life, I don't know of anything that will cause cold chills to run up and down my spine any more than seeing someone working on a well with the wire strewn about on the ground in the working area. I have never experienced this kind of wreck, but I can easily see what might happen if the pipe and pump were dropped and the wire got a wrap around someone.

There are several reasons that I do not advise using ordinary flexible plastic pipe for the water column on submersible pumps. This is true even if a safety cable is installed with the pump. (The safety cable is necessary because plastic pipe is prone to break, and you should have some way to pull the pump out of the well should this become necessary.)

First of all, the torque of the pump will in time weaken the pipe because each time it starts the pipe will flex slightly. This will in time also wear the insulation on the submersible wire, because as the pipe flexes, it acts as an abrasive. The end result is a short circuit. But in the event that the water in the well is corrosive enough that plastic pipe is preferred over steel pipe, then just be sure to use rigid plastic, thread and collared PVC.

I offer no guarantee for submersible pumps set on flexible plastic pipe, and for sure not in wells that are uncased. If a rock or cave falls out of the wall and wedges the pump, then you are in trouble. The only way to move the pump is up, and it will

72

usually come off where it joins the plastic pipe. If, however, the pump is set onto rigid plastic or steel pipe, there is a chance of moving the pump down a little — or turning it — to get it loose.

With a submersible pump system it is neither necessary nor desirable to house the control box and pressure tank directly over the well. In order to allow access for service of the well or the pump, it is far better to offset the tank and control house from the well, as it will usually accumulate an assortment of golf clubs, baby carriages, vacuum cleaners, and other miscellaneous equipment.

The water system should be on a separate circuit from the main power source, and it should have a fuseable disconnect switch near the pump control box. I am not satisfied with the motor being marked "lightning protected," and advise a lightning arrestor in the fuseable disconnect circuit or in the control box as added protection. Lightning is a major cause of submersible pump motor failure, and does not necessarily occur completely during the electrical storm. The motor, damaged due to lightning surges on the line, may continue to operate for an extended length of time before complete failure occurs.

A relief valve should be installed between the pump and the pressure tank. In the top of the tee that holds the water column is the logical place to put a relief valve in order to compensate for a block between the pump and the pressure tank. These blocks can occur because of a frozen pipe — which is usually the case — a stuck or improperly installed check valve, or points stuck closed on the pressure switch. As proof of the necessity for a relief valve, I offer the following incident. One time I received a long distance call from a friend who had rented one of his houses to a Mechanical Genius. The renter had been out of water for several days when my friend went to the well, bumped the pressure switch, and started the pump. Everything was back to normal for a few days, then the switch malfunctioned again. The renter went down and bumped the switch as my friend had done, but the pump did not start. The renter took the cover off the pressure switch, whittled a stick, then wedged the points closed on the switch. For sure this started the pump, and I have no way of knowing how far back toward the house the renter got before the pump blew the bottom out of the pressure tank, the tee off of the water column, and dropped the pump, pipe and wire to the bottom of the well.

73

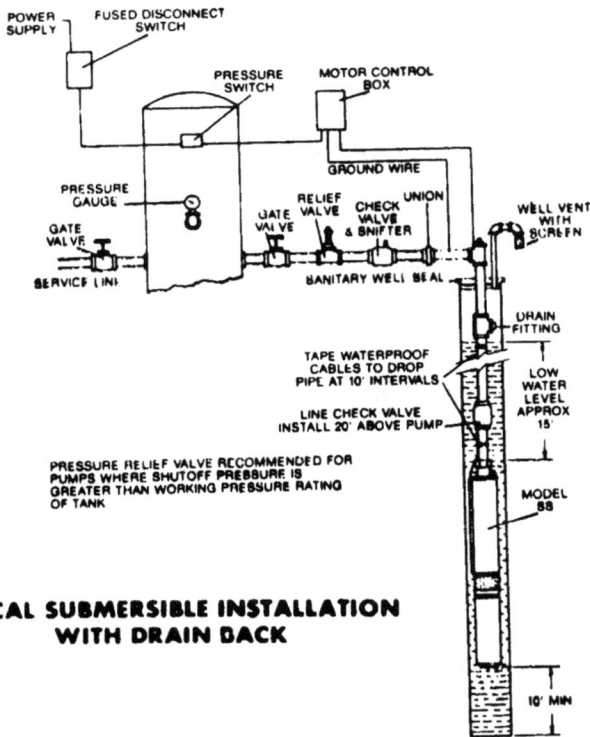

**TYPICAL SUBMERSIBLE INSTALLATION
WITH DRAIN BACK**

— Courtesy of Aermotor Division, Valley Pump

This was one instance where the relief valve would have been worthwhile. I would have given half the price of that fishing job to have been able to see that Matador take off when the tank blew up.

Your water system must be protected from freezing, and I most emphatically do not favor thermostatically controlled electrical devices such as heat tapes, heater, or light bulbs. It has been my experience that many times during severe cold weather, and especially in rural areas, there are many power failures. These, naturally, render such electrically controlled gadgets totally useless. It is much safer to wrap the pipe and fittings with weatherproof insulation, giving special attention to the one-fourth-inch nipple that is used to mount the pressure switch. I have made many service calls just to thaw out that particular nipple.

Another common problem of the pressure system is a water-logged pressure tank. This situation occurs when the air cushion in the top of the tank is not maintained, and the tank

74

becomes completely full of water. Each time a small amount of water is drawn from the tank, the pressure will decrease to the starting pressure of the pump, then immediately increase the pressure back to the cut-off point. If this starting and stopping of the pump is allowed to continue for an extended length of time, it will surely cause failure to either the control box or the motor, or possibly both.

If a standard pressure tank is used, then a bleed-back system should be installed in the water column. The bleed-back system consists of a fitting package with a tee the size of the water column. A bleed valve that inserts in the tee, a snifter valve, and an air volume release finish off the picture. Properly installed and maintained on submersible pumps, this is the best system that I know of to prevent water-logged pressure tanks. This is how it operates. The tee and the bleed valve are installed in the water column from six to eight feet below the check valve and the snifter valve. When the pump stops, the water drains out of the water column from the check valve to the bleed valve. That much of the water column fills with air, and then when the pump starts again it forces that much air into the pressure tank. When a surplus of air accumulates in the tank, it will force the float of the air volume control in the tank downward and release the surplus air. In this way the air cushion is maintained and the tank will not water-log. The most important part of this set-up is the snifter valve, which works like and resembles an automobile tire valve with one exception: there are one-quarter-inch pipe threads on the bottom end that screw into the check valve. The reason for having this valve is that it serves as an air vent so that the water column will drain to the bleed valve and fill with air. Now the monkey in the tree for this valve is that it comes with an air-tight cap screwed on the core end of the stem. The reason for the cap is still a mystery to me, so throw it away because no air can enter if the valve itself is capped. It is also possible that you might have to loosen the tension on the valve core so that air can be sucked in with a minimum of effort.

One more thing about this system. When the air volume control releases the surplus air there will be a hissing noise coming from the tank. This is as it should be, and no cause for alarm.

There are precharged pressure tanks on the market today

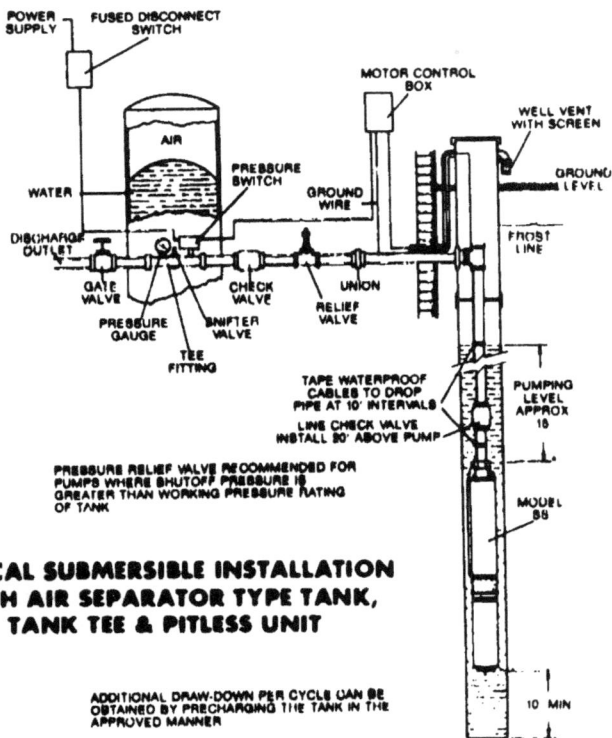

TYPICAL SUBMERSIBLE INSTALLATION WITH AIR SEPARATOR TYPE TANK, TANK TEE & PITLESS UNIT

— Courtesy of Aermotor Division, Valley Pump

that eliminate the need for the bleed-back system, and I certainly recommend using them. But a word of caution is advisable at this time: many times a standard pressure tank that has been equipped with a bleed-back system will be replaced with a precharged tank without the bleed-back valve being removed from the water column. The result is surplus air in the service line. To say the least, this is an aggravation to housewives and to all concerned. The remedy is to simply pull the water column and remove the bleed-back valve.

I have equipped many low yield wells, even those with a flow of less than a half gallon per minute, with liquid level controls, a submersible pump, and a storage tank. A liquid level control consists of a relay mounted near the control box and two electrodes suspended in the well by the electrode wire. The bottom electrode is set near the pump. The setting of the top electrode will depend on the static water level of the well, usu-

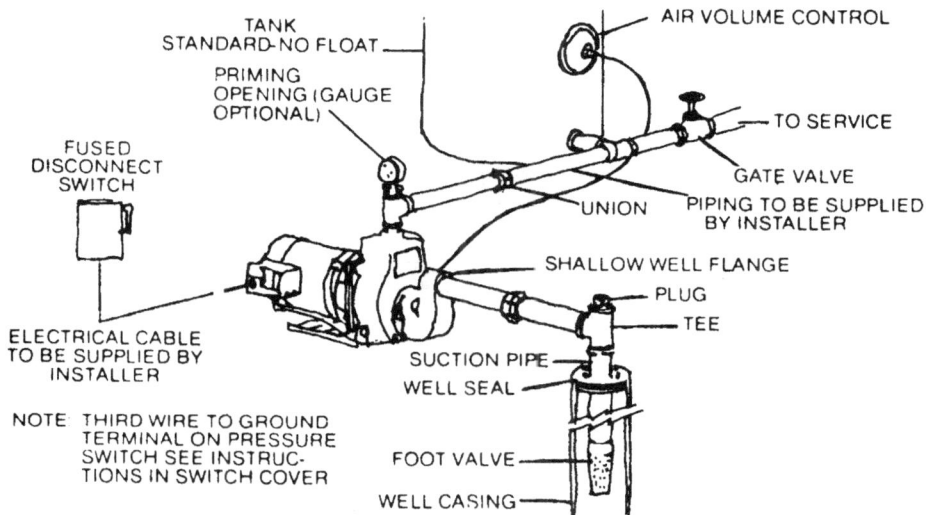

TANK STANDARD-NO FLOAT
PRIMING OPENING (GAUGE OPTIONAL)
AIR VOLUME CONTROL
TO SERVICE
GATE VALVE
FUSED DISCONNECT SWITCH
UNION
PIPING TO BE SUPPLIED BY INSTALLER
SHALLOW WELL FLANGE
PLUG
TEE
ELECTRICAL CABLE TO BE SUPPLIED BY INSTALLER
SUCTION PIPE
WELL SEAL
NOTE: THIRD WIRE TO GROUND TERMINAL ON PRESSURE SWITCH SEE INSTRUCTIONS IN SWITCH COVER
FOOT VALVE
WELL CASING

Jet and pump installation.

— Courtesy of Aermotor Division, Valley Pump

ally from five to ten feet above the lower electrode. The power to the liquid level control should come from a float switch mounted on top of the storage tank. When the tank is full of water the float will open the contacts of the switch, breaking the circuit to the liquid level control. The pump will not operate until the water — and float — lowers enough to close the points. When the points of the float switch are closed, power is supplied to the liquid level control electrodes. When both electrodes are submerged in water the pump will operate until the water level is lowered to the bottom electrode, at which time the circuit is broken and the pump stops. When the water level rises back to the top electrode, the circuit is restored and the pump begins to operate again, assuming that the float switch is still closed. With this kind of set-up, it's possible to get all of the water that the well will yield. A six by eight foot storage tank will hold about seventeen hundred gallons of water, and once the tank is filled a half gallon a minute well is usually sufficient to take care of the average household.

Now for the location of the storage tank. Unless a tall tank tower or a hill is available to provide for gravity flow, the tank should be set on a gravel base near the well. I say gravel instead of wood or concrete because the gravel will absorb the moisture that just naturally accumulates under the tank bottom. So placed, the storage tank will last much longer. When the storage tank is set up in this manner, an in-line pressure

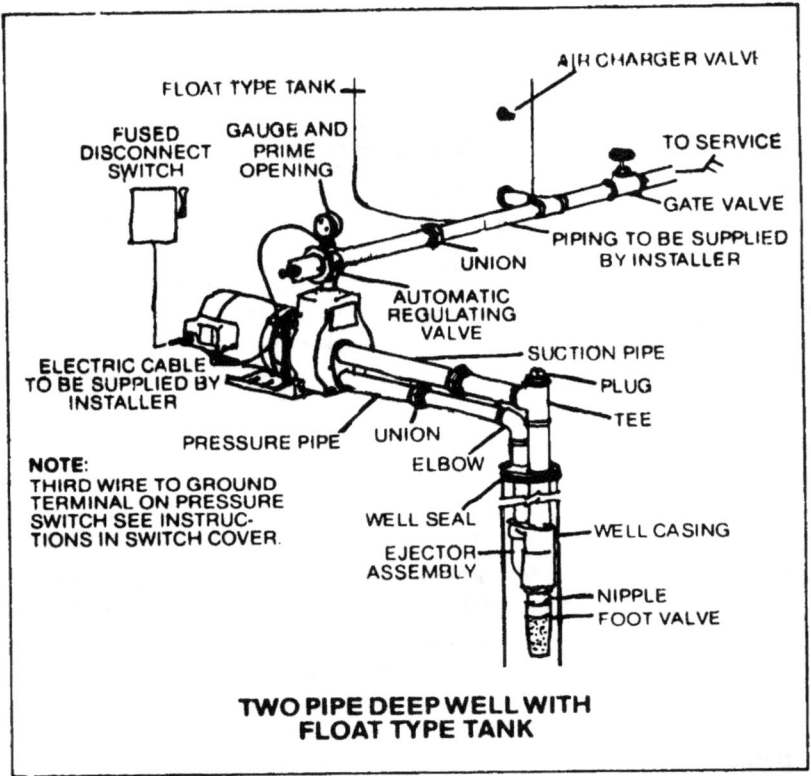

**TWO PIPE DEEP WELL WITH
FLOAT TYPE TANK**

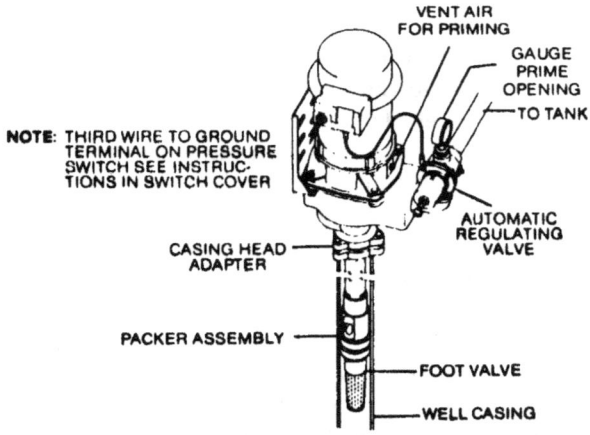

**DEEP WELL PACKER TYPE INSTALLATION
VERTICAL PUMP MOUNTING-CLOSE COUPLED**

— Courtesy of Aermotor Division, Valley Pump

system will be required in the service line. A good system is a one-half horsepower shallow well jet pump and an eighty-two gallon pressure tank with a check valve between the pumps and the storage tank. And this check valve is a must.

It is also a good idea to put a safety pressure switch in the system in place of the regular pressure switch on the pump. This arrangement is necessary because if the storage tank should happen to run dry, and the pressure drops below ten pounds, the points on the switch will not automatically close but will have to be closed manually with a lever on the switch. This arrangement prevents the pump from running dry.

It has not been my intention to degrade the use of jet pumps in the pressure system. I have seen many jet pumps installed that have worked well for twenty or thirty years, with a minimum of maintenance. For the most part these installations are on wells of one hundred feet, or less, in depth. Jet pumps may, however, be used for greater depth with the proper jet assembly and with additional stages on the pump. They may be installed horizontally, vertically, or in any combination. The horizontal pump may be offset from the well, while the vertical pump is mounted directly over the well. The combination may be mounted either way. The pump case should definitely be protected against freezing. The reason that I specifically refer to the pump case is that the motor will not freeze and should not be insulated. To do so may cause the motor to overheat and kick out the overload protector.

Regarding the function of jet pumps, over shallow well installations (20 feet or less) only one drop pipe with a foot valve is required. Deeper installations require two pipes, a jet assembly, and a foot valve. The nozzle size in the jet assembly will be different for each twenty additional feet of well depth, so be sure and use the correct one for the well you are installing the pump in. Usually the drop pipes will be different in size, the suction pipe being the next size larger than the pressure pipe. Some pumps, however, will use the same size pipe for both drop pipes. Many times in such cases the pump cannot build up enough pressure to operate the pressure switch. For such situations, simply change the pipes at the pump, as they have probably been put in backward.

In wells of sufficient diameter the foot valve, jet assembly, and the drop are installed side by side. On wells that are less than four inches in diameter, one drop pipe may be installed inside the other, with a jet packer assembly and foot valve on the inside pipe, and an adaptor on the pump.

The side-by-side drop pipe installation may be made with either steel or plastic pipe. When you use plastic pipe, get the best NSF (National Sanitary Foundation) pipe that is available. For the plastic pipe adaptors, I recommend metal, either galvanized or bronze because the plastic or Teflon adaptors will usually be broken should it become necessary to loosen them. The adaptor clamps should be made of stainless steel.

Most of the dissatisfaction with jet pumps, I believe, stems from the loss of prime. There are different reasons why this may occur. In order to operate, the pump must be completely primed. By this I mean that the drop pipes and pump case must be completely filled with water and there must be absolutely no air. When the pump starts, if there is no restriction of the discharge many times the prime will be lost. To put it another way, the pump discharges more water than it picks up, thus losing the prime. For this reason an automatic spring loaded control valve is used between the pump and the pressure tank. The same procedures may be accomplished manually with a gate valve by restricting the flow and maintaining ten or twelve pounds of pressure on the pump until this amount of pressure is in the tank. Then the valve may be opened to capacity.

Another reason for loss of prime may be because you have pumped the static water level down to the foot valve, and last but by no means least is a partially clogged nozzle in the jet assembly. Many times the clogged nozzle may be remedied by completely priming the pump, restricting the discharge flow, and starting the pump while at the same time giving the pump and drop pipes a good schoolboy shaking. If this fails to dislodge the blockage, then the jet assembly will have to be pulled and the nozzle cleaned.

Going back to the basic pump: I prefer the vertical installation because in my opinion there is less wear on the shaft bearing and the seal. And one more thing — although it's not a common practice, the jet pump may be powered with a gasoline engine instead of an electric motor.

The pump-jacks that I intend to relate to are used mostly to supplement the windmill in emergencies, however, there are instances of permanent installation. Pump-jacks are usually

OPEN TOP CYLINDER

CLOSED TOP CYLINDER

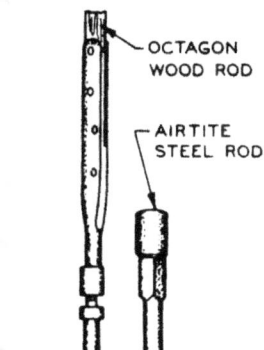

- OCTAGON WOOD ROD
- AIRTITE STEEL ROD
- PIPE COUPLING
- WELL PIPE
- STEEL ROD
- ROD COUPLING
- WOOD ROD OR AIRTITE ROD

PRINCIPLE OF CYLINDER OPERATION
Pump rod of wood or steel connects plunger of pump to valves in cylinder and brings about actual pumping action

There is a check valve at bottom of cylinder and a similar valve in the plunger. Water flows into cylinder through check valve while plunger is making its up-stroke. On down stroke, water is held in cylinder by check valve and plunger descends to bottom while water passes through valve in plunger.

On next up stroke, valve in plunger closes and water above it is raised up into drop pipe. At same time check valve opens and cylinder fills with water again. With each up stroke of the pump, the plunger forces a cylinder full of water into drop pipe and out to discharge tank.

SELECTION OF CYLINDER
...en top cylinder is recommended ...ere it can be used. Inside diame-...of the drop pipe is slightly larger ...n inside diameter of cylinder. ...s permits lowering or removal of ...nger and check valve through ...o pipe.

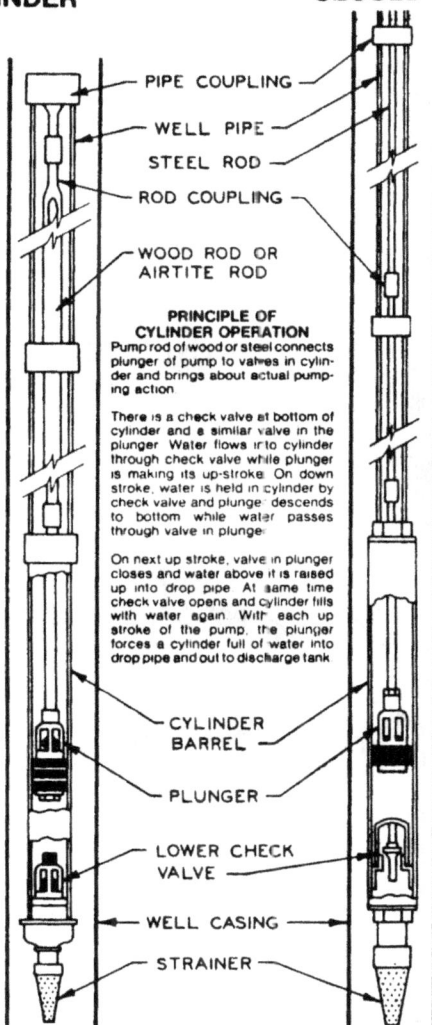

VALVE ASSEMBLIES

BALL

SPOOL

GRAVITY POPPET

- CYLINDER BARREL
- PLUNGER
- LOWER CHECK VALVE
- WELL CASING
- STRAINER

ROD COUPLINGS

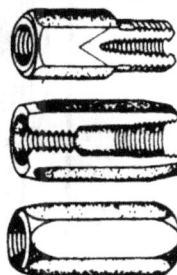

STEEL SUBSTITUTES

— **Courtesy of Aermotor Division, Valley Pump**

powered with gasoline, diesel, or butane engines. One of the most common problems encountered with a pump-jack is lugging the engine and causing the valves to burn. These engines have to run fast enough to oil themselves and if the pump-jack is making too many strokes per minute the engine should be equipped with a gear reduction or pulley adjustment rather than slowing down the engine. Discard the notion that the

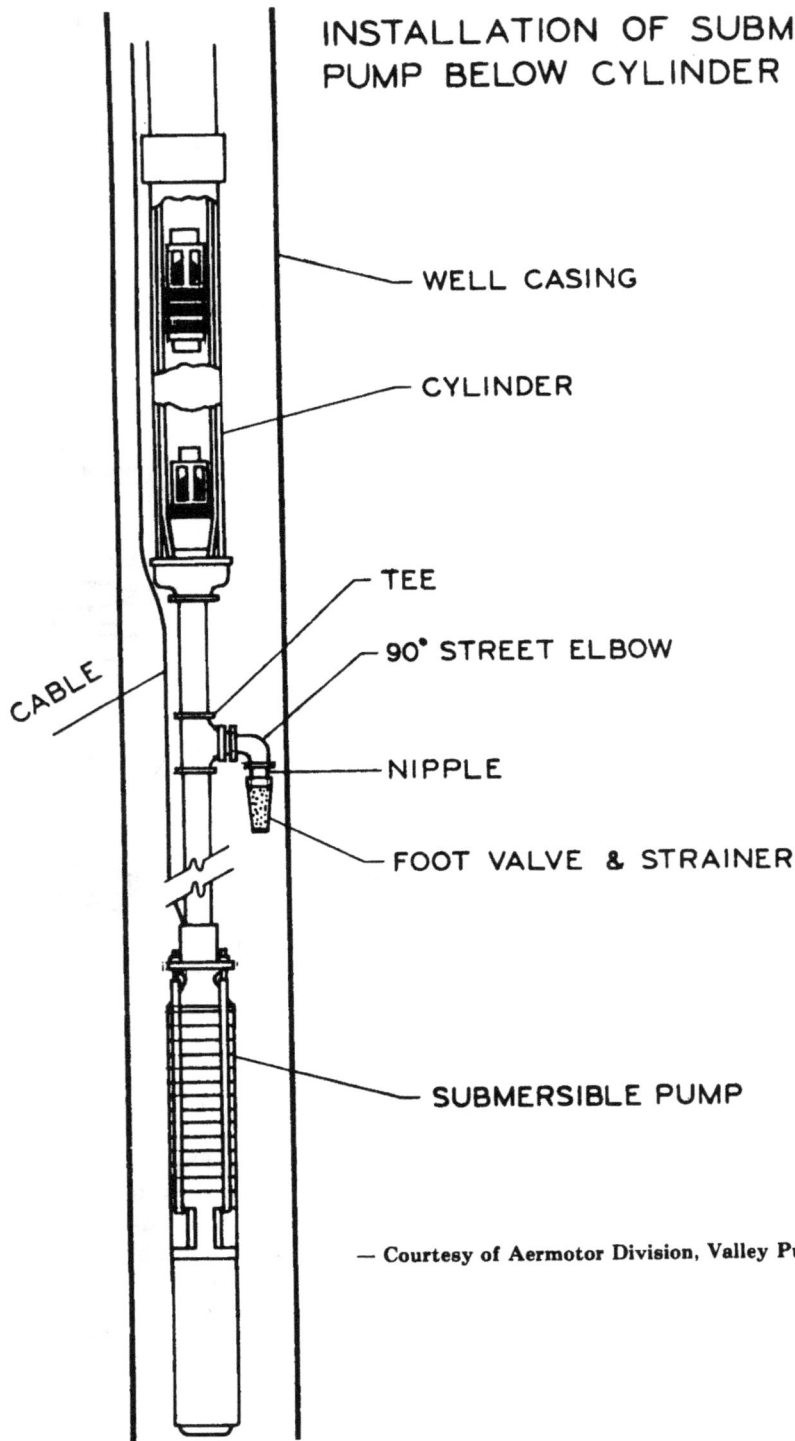

INSTALLATION OF SUBMERSIBLE PUMP BELOW CYLINDER PUMP

WELL CASING

CYLINDER

TEE

90° STREET ELBOW

NIPPLE

FOOT VALVE & STRAINER

CABLE

SUBMERSIBLE PUMP

— Courtesy of Aermotor Division, Valley Pump

faster the jack runs the better. The number of strokes per minute depends on the depth of the well and the length of the stroke. Running a jack too fast will pop the rod and most certainly cause trouble. It is also advisable that the stroke of the pump-jack not be longer than the stroke of the windmill to avoid trouble in the cylinder. Most pump-jack installations are near ground level and will require a stuffing box. Operating a pump-jack with an electric motor is, I believe, a waste of time, money and energy. As an alternative I would suggest installing a submersible pump on the bottom of the windmill cylinder. An installation such as this will eliminate not only the pump-jack and motor, but the need to take the windmill loose from the sucker rod when using the pump-jack. It will also pump a lot more water. In situations in remote areas, where there are several windmills that may require auxiliary pumping, I would suggest the submersible pump on the bottom of the cylinder and a good portable 220 and 110 volt generator (which can also be used to operate power tools) in remote areas.

When installing the submersible pump below the cylinder, without an offset valve, the windmill will pump through the submersible and the pumping action of the windmill will rotate the pump. It is debatable whether this situation is good or bad, but it may be remedied by an offset foot valve between the cylinder and the submersible pump. To install the foot valve between the cylinder start with a close nipple in the bottom of the cylinder, a tee, another close nipple, then the pump. Put a street ell in the tee and install the foot valve in it. This will enable the windmill to pump through the foot valve instead of the pump.

One problem that could possibly be encountered in this type installation might be a crooked well bore less than six inches in diameter. The water-column for the windmill being larger and much more rigid than the water-column (usually one inch or one-and-one-fourth inch) intended for the submersible may cause the pump to bind in the crook in the well bore. On any submersible pump installation extreme care must be taken not to damage the insulation on the electric motor lead wires.

VI

OWNER MAINTENANCE AND TROUBLESHOOTING

Maintenance will be required whether you own one wind-mill or a dozen — or more. I guess that perhaps the best way to go about this is to write about the owner who has several wind-mills, and plans to keep them in operation himself. Much of what applies to him will apply to the owner of only one mill.

The following is a true story that reads a lot like fiction. I'll just throw it in here for whatever it's worth.

Near Silver City, New Mexico, there was an old sixteen foot Samson windmill on a forty foot wide-spread tower. It was located on a fairly large ranch that has several mills, all deep wells. Most of these were drilled during World War II when casing was pretty hard to come by and consequently they were mostly uncased. That in itself adds to all the other problems, and one morning the ranch foreman discovered that the one mill wasn't working.

Now, it's an established fact that cowboys have a certain amount of allergy to any kind of work that can't be done from a horse's back and that working on a windmill is up close to the top of cowboy's list of dislikes.

The first mistake that the foreman made that morning was when he told two of his cowboys to turn their horses out, take the pickup and some tools, and go fix that old windmill. There was a fairly complete set of windmill tools scattered around th' hay barn, saddle room, bunk house, garage, and front yard. Along with these, Slim and Chunky loaded th' wire stretchers, two pipe wrenches, a saddle rope, and the pliers out of th' boss's Cadillac, shoved the pickup off the hill to get it started, and took off for that ailing windmill.

A sixteen foot mill over a hole in the ground should tell you something about how far toward China it is to water, and the sucker rod tied to that windmill had to go all the way. And it stands to reason that th' more there is, th' heavier it is.

When Slim and Chunky pulled up to that windmill, there was an occasional gentle breeze blowing, enough to turn th' mill over kinda slow and easy. Right off, Chunky went over and cranked the furl winch to shut it down, but nothing happened — the winch wire was broken. Even though, Slim decided that there was no problem that between the two of them they couldn't take care of.

On this old mill the pump rod had had th' threads stripped a time or two at the cross-head. It had been cut off and re-threaded and was now too short for the swivel to clear the pull-out assembly on the up-stroke. But on the other hand, what did that have to do with a cow? So Slim, spurs and all, climbed th' tower with a piece of barbed wire, spliced the furl wire, an' hollered down to Chunky to cut 'er off. Chunky almost had the wheel and tail together before th' swivel came up and hit th' pullout assembly and broke the furl wire again. At that point Slim decided to hell with it — they didn't need to cut it off anyway.

From here on out what happened would make a pretty fair plot for a late night horror movie. In a forty foot wide-spread tower th' span between th' girts fifteen feet above th' ground is pretty long for a rotten two by twelve plank, and that's where th' sucker rod was bolted to the red rod. That's also where that two by twelve had been for the last twenty years.

Well, Slim had used up his brainstorm on how to fix the winch wire. Now it was Chunky's turn to use his noggin for somethin' besides to keep his ears apart. "Hell, Slim," he says, "ain't nothin' to it, you git up there on that two by twelve and take th' nuts off them bolts an' take all th' bolts out but one. I'll get that block of salt out there to stand on so's I can reach the standpipe. When she comes to the bottom of th' stroke, just before she starts up, pull the last bolt and I'll hold th' rod." Slim looked at Chunky from under th' brim of his hat and asked, "Think you can hold 'er?" "Hell, man, can a bullfrog jump?" replied Chunky. Well, it seems that somewhere back down th' line Slim had had some experience with frogs, for he told Chunky, "Suits you, suits me!"

Slim went after the boss's pliers and Chunky went after th' salt block. Next thing Slim's on th' two by twelve, which now had no more than two or three inches of sag in it, and had th' bolts all out but one. And Chunky's all set on th' salt block. Th' mill's still turning kinda slow, the rod goin' up and down sorta easy like. "Next time," hollered Chunky, "turn 'er loose." Th' next time Slim did just that — he pulled th' last bolt and th' rod and Chunky headed for China. They didn't get far, though, when Chunky's hands wedged in th' standpipe. There he was, stompin' on th' salt block and screamin' "I'm hung." "Turn 'er loose!" hollered Slim. "I can't, dammit, I'm hung." "Hurry, lift it up," cried Chunky. So Slim grabbed a good hold with both hands, flexed his legs, an' heaved. You guessed it. That old plank broke, and Slim, spurs and all, came down astraddle Chunky's neck. Chunky claims he never would'a got loose if it hadn't been for Slim's spurs. As it was, he came away with both hands and all ten fingers, but minus a lot of his hide. And now he has a lot better idea about th' weight of five hundred feet of number two sucker rod!

All these mistakes made on this job due to carelessness, inexperience, laziness, day-dreaming, or whatever, are in no way uncommon. Now I'm going back over this story to try to point out some of their errors.

I think the first mistake made was when the foreman hired Slim and Chunky. He failed to explain, or at least didn't get it across, that water, not the boss's Cadillac, was the most important item on the ranch, and that it was going to be up to them to see that it was where a cow would get to it without going five hundred feet underground.

The next mistake I see is the tools. Assuming that there were enough on the ranch to do the job right, it might have taken until noon to find them. It would have been much better to have all the tools under the kitchen table, or at least somewhere in a group. However, pretending to be as wise as I am, I would offer as an alternate solution a trailer equipped with a weatherproof tool box and a pipe rack. All the tools should be kept in their proper place in the trailer, and anyone using one and not returning it would commit the unpardonable sin. Any discipline, short of hanging, or the firing squad, should not be considered too severe.

As long as I'm only suggesting, I may as well list at least some of the tools that should be kept handy — preferably in the trailer. They are: safety belts, acetylene cutting torch, welding goggles, welding tips and rod, cables and clamps for pulling pipe and sucker rod, three cable sheave snatch blocks, ropes, chains, elevators for pipe and sucker rods, three twenty-four-inch chain tongs, two eighteen-inch standard pipe wrenches, crescent wrenches, pliers, vise-grips, hammers, screwdrivers, punches, chisels, pipe vises, pipe dies, windmill oil, an assortment of cup-leathers, sucker rod ends, copper rivets, bolts, and pipe fittings. The list could go on and on from here, but these will usually do the job.

Putting together this set of windmill tools will not come cheap, and may even necessitate cutting the stitching in the boss's boot tops from five rows down to three. But in case you are looking for a bargain, I'll offer this bit of advice from now until thirty minutes after sundown, and at a twenty percent discount: buy the boss a pair of brogan shoes and hock the Cadillac. Oh yes, I almost forgot. If there's any loot left, get a new battery for the truck.

I would say that the next mistake was when Chunky went to turn the mill off without giving any thought as to why it might not be pumping, or how long it had been since it had pumped. He could have looked to see how much water was in the trough and tank, and whether or not the end of the lead-off pipe was damp. It's possible that the mill might have pumped some water with more wind — if so then the cup leathers were probably worn and should have been replaced. While Chunky was checking topside, Slim could have been listening for the check valves to see if they were clicking. If he couldn't hear them a sharp rap on the water column — on the up stroke — might have put them to working. If they were working, and there was still no water in the standpipe, there could have been a hole in the water column. If that had been the case, they could have forgotten about the wire stretchers and hunted something else — bigger and better tools.

When Chunky took hold of the furl winch handle to turn the mill off, he was drawing to an inside straight, and whoever had cut the pump rod off and rethreaded it, had stacked the deck against him. In spite of these problems, Chunky's luck had returned when the furl wire broke instead of the winch handle getting loose and knocking him on his quatro-cinco. Among other things that could have happened had the furl wire been replaced with cable or chain was that sooner or later the main shaft could have broken and the wheel would have blown off. Other possibilities are that the gears could have been stripped, the pull-out chain could have been broken, the threads could have been stripped in the cross-head or the pump rod end (maybe both), or the swivel could have been pulled off. Why? Because, when furling a mill like this Samson, care must be taken to see that the pump rod swivel will clear the bottom of the pull-out sleeve at the top of the stroke. In this case — as in many — the pump rod had been cut off and rethreaded and was thus too short to clear the bottom of the pull out sleeve. When you replace a furl wire with a cable or chain, you have a lot stronger system, and you can cause the sleeve and pump rod to collide at the top of the stroke.

The best and safest way to furl a mill such as this when you are going to work on it is to tie the wheel and tail together. There is a right way and a wrong way to do this. First and foremost, the wheel must be kept from turning. This can be done by throwing a rope over the tail, then someone on the ground pulls it into the wind, on the furling side, until the wheel stops. The rope, chain, or whatever is to be used to tie the wheel and tail together, must be long enough to tie to the outside band around the wheel arm in two places, thus forming a "V." Tying the wheel to the tail — straight across with one tie — just will not do. In making the tie, first double half-hitch the rope or chain in the middle around the tail-bone, as far out as can be reached. Next, tie the rope or chain to the outside band around a wheel arm at the bottom of the wheel. Then turn the wheel backward until the rope or chain is at the top. This will completely furl the mill. Now tie the wheel to the other end of the rope in the same manner as before, thus forming a "V." Make sure that this tie is strong enough to keep the wheel from turning either way — not even a little bit. The mill is now free to rotate with the wind and be left for long periods of time with no damage. But if there is any play at all in the tie, watch out, because sooner or later the action of the wind and the constant flexing of the cable or rope will cause it to break, and then you might have problems.

Another method that is commonly used is to tie the wheel to the tower. This may be all right if the wind isn't too strong and if the tie is made completely around the tower, not just to one side. But don't use it for an extended length of time.

The next mistake — not our cowboy's fault — appears where the sucker rod was bolted to the red rod. Instead of this connection being made fifteen feet up in the tower, it should have been made where it could have been reached from the ground, thus eliminating the use of the rotten two by twelve and all the errors that naturally went with it.

Not being as well versed about bullfrogs as Chunky and Slim claimed to be, I'll leave them with the rest of their mistakes and experiences and go on with windmill maintenance.

MAINTENANCE

While it is true that the revolutions per minute (rpm) of a windmill are generally much fewer than most machines, it is well to remember that this particular mechanism may work twenty-four hours a day, seven days a week. That totals a lot of rpm's. Therefore, the life of a windmill is dependent — to a large extent — on proper lubrication.

The design of the Aermotor oiling system, in my opinion, is one of the best, if not the best. I do not believe, however, that Aermotor engineers consider it necessary that the helmet be ventilated with bullet holes. My main objection to installing air vents in this manner is that they allow dust to enter the motor. In drier times, blowing dust accumulates on the outside of the helmet, then when it rains the water and dust — or mud if you will — washes into the motor case and forms a sludge. Sludge is also formed by using detergent oils instead of the non-detergent variety. I will be among the first to admit, though, that any kind of oil is better than none at all.

Sludge is probably the number one enemy in any oiling system, and especially the one that Aermotor devised. Let me explain. The cogs on the large gears pick up oil from the bottom of the motor case and carry it up to the small gears. On the wheel-side of the small gears is a spout washer held against one of the gears by a spring around the main shaft between the gear and the main shaft bearing. The spout washer picks up the oil from the small gear and deposits it on the top of the main shaft. This oil lubricates the main shaft bearing and the

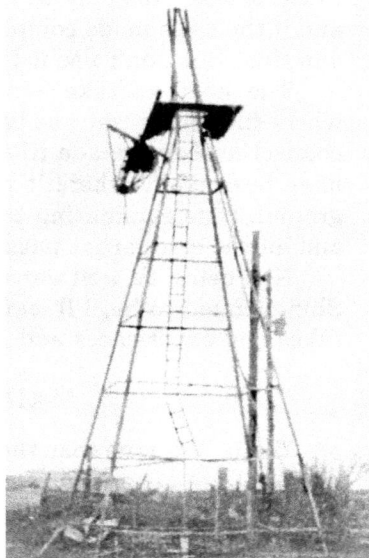

Lack of grease may have caused these disasters.

excess is collected by an oil collector inserted through the case and slot in the outside end of the main shaft bearing. The oil then returns through a channel in the bottom of the main shaft case to the motor case. If this route becomes plugged — and it's usually sludge that does it — the oil will run out through the hub. You can clean this channel with a piece of welding rod or some kind of stiff wire.

The pipe-plug in the hub is put there for the sole purpose of allowing access for removing the oil collector in the event that the main shaft has to be pulled. This plug has nothing whatever to do with oiling the mill.

This pretty well takes care of oiling the bottom part of the motor: now for the top part. An oil ring is fastened on the side of the pump-rod yoke in a manner so it will turn some when it touches the top of the large gear at the bottom of the windmill stroke. When it touches, it picks up oil that in turn drips on the shaft that goes through the pitman ends, the pump rod yoke, and the guide wheel. I need to emphasize at this point the im-

90

Instructions for Changing the Stroke

1. The mill was assembled on the long stroke at the factory.

2. Capacity tables are based on the long stroke of the mill.

3. Setting the mill on the short stroke decreases the capacity by 25%.

4. Setting on the short stroke may accomodate the present cylinder in the well if the barrel length is not sufficient to allow clearance of the long stroke.

5. On the short stroke the mill will run in a lighter wind.

To Change This Mill to the Short Stroke

1. Turn the large gears until the pitmen are near the bottom of the stroke. (If the pump rod is under load, support the pump rod yoke with a wood block as shown in Figure 16. A piece the size of the pump pole will do.)

2. Remove cotter pin at one end of No. 522 shaft and push the shaft out from the pitmen, yoke and guide wheel.

3. Turn large gears until lower ends of pitmen are above rim of gear case. Remove cotter pins and No. 622 bolts, and take both pitmen off the gears.

4. Turn the large gears about one-half revolution, or until short stroke bosses are above rim of gear case, so that pitmen can be placed on them. These bosses on the large gears are the ones that are closest to the center of the gears.

5. Place both pitmen on the short stroke bosses, making sure to replace the No. 622 Bolts and the cotter pins that hold the lower ends of the pitmen on the large gears.

6. Turn gears until pitmen are near bottom of stroke, as in Figure 17.

7. Replace No. 522 shaft thru **lower holes in upper ends of both pitmen**, thru pump rod yoke and guide wheel. Be sure to replace cotter pin in No. 522 shaft.

8. After changing the stroke, be sure that the oil ring rolls on the large gear at the bottom of the stroke. Turn the wheel to be sure everything works freely before putting the mill in operation.

IMPORTANT:
WHEN CHANGING STROKE, BOTH ENDS OF PITMEN ARMS MUST BE CHANGED AS SHOWN.

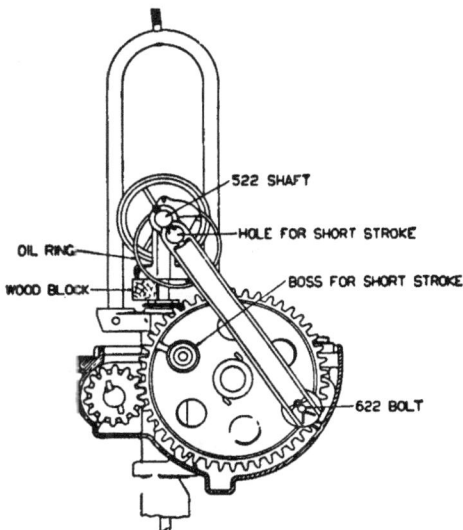

522 SHAFT
HOLE FOR SHORT STROKE
OIL RING
BOSS FOR SHORT STROKE
WOOD BLOCK
622 BOLT

MILL SET ON LONG STROKE

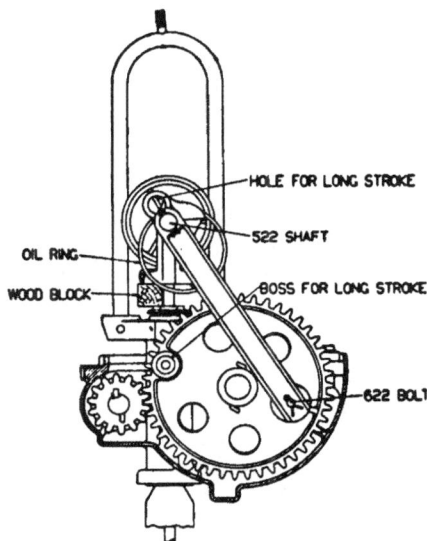

HOLE FOR LONG STROKE
522 SHAFT
OIL RING
BOSS FOR LONG STROKE
WOOD BLOCK
622 BOLT

MILL SET ON SHORT STROKE

— Courtesy of Aermotor Division, Valley Pump

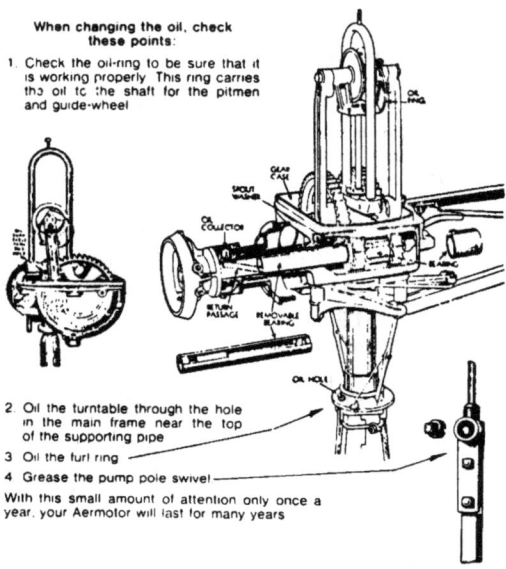

When changing the oil, check
these points:

1. Check the oil-ring to be sure that it is working properly. This ring carries the oil to the shaft for the pitmen and guide-wheel

2. Oil the turntable through the hole in the main frame near the top of the supporting pipe

3. Oil the furl ring

4. Grease the pump pole swivel

With this small amount of attention only once a year, your Aermotor will last for many years

— Courtesy of Aermotor Division, Valley Pump

portance of this oil ring. It is the only way that the top part of the motor is going to get any lubrication. If the oil ring becomes bent or broken it should most certainly be replaced.

The most common mistake made by amateur windmill repairmen is splitting the stroke of the pitmans. The stroke is split by changing only one end of the pitmans when altering the stroke of the Aermotor windmill. This should always be avoided. The top hole of the pitman mechanism matches the outside boss on the gears, and the bottom hole is supposed to match the inside boss. If only one end is changed, this "splits the stroke," and the oil ring will not be able to pick up oil.

As for the oil, in most cases it should be non-detergent, and no more than ten weight. It must be checked periodically, and should be changed every year. It is also a good practice every third or fourth year to give the motor a good bath with solvent or kerosene.

That takes cares of the self-oiling part of the Aermotor windmill. Now for the parts that need to be manually lubricated each time the oil is changed: they include the mastpipe and turntable, the furl ring, the pump rod swivel, and the rub blocks. The only way to oil the mast pipe and turntable is with a squirt can through a small hole in the case that leads to the mast pipe. The furl ring has an opening with a one-fourth-inch pipe-plug inserted in it. Remove the plug and oil the ring, or better yet throw the plug into the water tank and replace it with a zert fitting and use a grease gun. Next, pack the pump rod swivel with grease, and finish by smearing a little on the rub blocks. With this, the lubrication act is complete.

What other maintenance is there? I suppose that it goes without saying, but I will say it anyway, that you should make sure that all bolts are in place and tight. If a hole is observed that even looks like it might require a bolt you should, without the slightest hesitation, proceed to "put 'er in." At the time of the yearly oil change, remove the helmet and hang it on the tail. This way you can get a clear view of and good access to the motor. Check the main-shaft bearing between the small gears and see that the knob that holds it in place — by a slot in the case — has not worn off, allowing the bearing to turn with the shaft. Check the operation of the spout washer, the oil ring, and the large gear bearing bar. Also check for slack in both the bottom and top of the pitmans, and check for clearance between the pump rod yoke and the mastpipe lock nut.

Now for maintenance on the tail and the wheel. The most likely trouble spot on the tail will be the tail-bone pivot bolt. Check this carefully for extensive wear, and replace it if necessary. Check all the wheel bolts and nuts. A very important point here is where the wheel arms screw into the hub — check carefully for any slack. This may be done by turning the wheel while at the same time pushing in and out on it. If you detect any slack at all in one or more of the arms, the threads are either stripped or broken. That particular arm should be replaced.

Assume that the bearing between the small gears needs to be replaced, a common maintenance problem. These bearings are listed in the manufacturers parts list and in Aermotor's extensive advertising as "replaceable babbit bearings." You could well gain the impression that their replacement is a task for Mamma and the kids, but unless Mamma has more windmill experience than most mothers, and the kids are at least half grown, you may have overmatched Mamma. (No reflection on the ERA intended.)

If this bearing is badly worn, you might as well plan on replacing the main bearing also. Additionally, you had better order a new spout washer too, because you will probably break it when you pull the main shaft.

First, disassemble the wheel. After this is done—right now before going any farther—turn the hub to the top, take out the pipe-plug, and remove the oil collector with a screwdriver or whatever it might take to unscrew it. The main shaft definitely cannot be pulled without removing the oil collector. Next remove the counter-sunk plug opposite the end of the shaft and take out the bent pin in the shaft end that holds the gear key in place. There is no way that I know of for pulling this gear key, so the shaft has to be pulled through the bearings and gears and off the key. Many times, this takes some doing!

After you have put in a new small gear bearing and main bearing, reassemble the shaft the same way you took it apart. It is a good idea to mark the first gear in some manner so that it may be aligned with the key-way in the shaft. This will make the whole thing easier to put back together. So much for the main shaft.

Replacing the large gear bearing doesn't present so many problems. Take the top end of the pitmans loose, remove the bearing bar, then take the big gear assembly out. Now drive out one of the gear shaft pins, take off a gear, and remove the bearing. Most people will simply put in a replacement bearing, but if babbit and experience are available, this bearing may be rebabbited. Also, if the bottom of the pitmans is worn, now is the time to rebabbit or replace them.

There is the possibility that on mills with years of service the combination of wear on the bearings, pitman ends, and turntable may be such that the pump-rod yoke may bump the mast pipe locknut at the bottom of the stroke. This condition may be taken care of by installing a turntable shim, called a split-washer, between the turntable and the motor. The split-washer is notched and may be installed by raising the mill an inch or two and placing each half around the mast pipe on top of the turntable. I have done this many times by taking the locknut off the top of the mast pipe and raising the motor slowly with the pump rod.

This operation merits a stern warning that the pitmans or bearing bar can and possibly will be broken, because we are stressing the mill exactly opposite to the way it was designed. In other words, we are going to push instead of pull, and for the next twenty minutes I offer no guarantee of any kind. But if you aren't afraid, then I'm not, so here goes.

Turn the mill all the way to the top of the stroke, tie the pump rod swivel securely to two sides of the tower, take the lock nut out of the top of the mast pipe, then carefully turn the wheel BACKWARD — and I do mean backward — so the pitmans will come straight down instead of at an angle. If the rigging holds, the mill will raise itself. Just in case somewhere along the line you lost your nerve, then put up a gin pole and raise the motor with that.

Other than protecting the water column against freezing, which I will cover later, I think this pretty well takes care of maintenance on the Aermotor windmill.

Many of the things you have to do to keep an Aermotor running right hold true for other makes and models. However, there are differences in oiling the top part of the motor, the tower cap and turntable bearing. Mechanically, the furling and braking assembly, the clamp and guide for the pull-out sleeve, the spider, hub bolts, and hub keys are also different.

Incidentally I would list the inside hub key as the number one villian for other than Aermotor mills. The most highly stressed point of these mills is the inside hub which carries ninety percent of the load. The hub is attached to the main shaft only by a hub key, and if this key should become loose, vibration will likely cause it to work out. This puts one hundred percent of the load where ten percent was intended, namely the outside hub. Disaster is sure to follow.

The ways to oil the top part of the motor are many and varied on the other kinds of windmills. Regardless of what other brand you might have, oil it yearly as you would an Aermotor, and make sure to check to insure that the oiling system is in working order.

On many of these windmills, the tower-cap holds the top of the tower together and provides a base for the turntable bearing. The entire weight of the mill and it's load rotates on this bearing. Failure to grease it will not only result in a ruined bearing but may also cause the tower cap to break — another disaster.

The furling and braking system on most of these mills operates with a pull-out sleeve and tail chain which go inside the mast pipe along with the pump rod. The first to break is usually the tail chain, and it will probably happen inside the mast pipe. If it does, be sure and get all the pieces of chain out of the mast pipe, because any that are left might serve as a wedge and lock the pump rod on the down stroke. That, of course, means big trouble.

If the brake band becomes worn — again on brands other than Aermotor — to the extent that it will not keep the wheel from turning, or if it is anywhere near breaking in two, then by all means discard it. Should it break, it's possible for a rotating wheel arm to pick it up and snap the brake band bracket off of the motor case, sometimes taking part of the case with it.

Attention should also be given to excessive wear in the spider. The spider is bolted to the four tower posts at the bottom of the mast pipe. Its purpose is to center the mastpipe in the tower, and keep the mill level and plumb. When and if the spider becomes worn or broken, there is undue stress on the turntable bearing and the tower cap. If there is enough play the wheel may be able to hit the tower as it rotates on the turntable. Here again, a wreck is sure to follow.

On these other mills the pull-out sleeve consists of a short length of pipe, actually a sleeve around the pump rod, whose length and size depends on the mill size. This sleeve is equipped with a swivel at the bottom and an arm that extends through two tower posts to hold it in line with the winch wire. Above the swivel is a strap welded to the sleeve. This strap is an inch or two longer than the sleeve and fastens to the tail chain. There is a clamp and guide bolted to the bottom of the mast pipe to insure that the sleeve and tail chain rotate with the mill. If the strap and guide become worn enough to allow the sleeve to turn inside the clamp, then the entire assembly should be repaired or replaced.

Last but by no means least, allow no bird nests to be built or at least remain in the tower. This is especially important if no protection has been provided (a finger-gitter) for the water column. Crows are hatched as thieves, and will steal anything they can fly with and hide in their nest. Their loot may include nuts, bolts, nails, bright colored rocks, bones, calf horns, and other objects too numerous to mention. The wind, and the vibration of the windmill, will often dislodge these accumulated trophies, and one may fall into the water column. When this happens, the best that can be hoped for is a ruined cylinder.

The roots of all these trees are "water hunters," and not unlike many big game hunters they follow the line of least resistance. Many times, that line is the bore of the well. I have cleaned roots, or at least attempted to clean them, from a depth of as much as eighty feet when the closest tree was one hundred feet from the wellhead. I would hesitate to venture a guess as to how long it took those roots to go that far but know that it was less than twenty years.

In most cases the roots are more abundant near the surface of the well, and quite possibly it may be best not to disturb them. However, if they become thick enough it may be impossible to get the cylinder or pump through the mass. Then, of course they'll have to be removed. At times like these I have often considered manufacturing root wigs for bald-headed hippies. But this is another business itself, and government rules and regulations could, and probably would, prohibit such practice. So — back to water wells.

One way to go about removing the roots is to use a fairly heavy bar with prongs welded on the side. Lower it into the well on a pipe or a piece of sucker rod instead of cable so that it may be turned to loosen the roots. The fewer passes that can be made, and still get an opening, the better. The reason I say fewer passes is because with each pass a few pieces of root will fall into the water. And the more pieces of root in the bottom of your well, the more trouble you will have. Roots — as well as sand and silt — require a cylinder strainer. Again, ball valves are recommended.

TROUBLE SHOOTING

When a windmill is running and everything above ground is in proper working order, yet no water is being pumped, there can be several reasons for the difficulty. Many times, instead of landing on it like a hen in a pile of horse manure, and tearing it all apart, you may save a lot of time and work by taking a few minutes to locate the problem. Just consider yourself a "water well doctor," but don't even consider sending a bill that may compare in amount to that of an MD.

The first thing I recommend is to use a pipe wrench or a short piece of pipe as a stethoscope, and while the mill is running, listen for the valves. (Just lay one end of the pipe or wrench against the water column, the other against your ear,

Replacement leathers.

and you're in business.) If the ball or spool in the stationary (bottom) valve closes when the sucker rod is at the top of the stroke, it will click, a ball sounding much louder than a spool. As the rod goes down, the ball or spool may be heard to rattle, and the click will come at the bottom of the stroke. If a click cannot be heard for the bottom valve, then it is either fouled or the cup leathers are worn to the extent that they will not pull it open. Many times a sharp rap on the water column with a sledge hammer or similar object will jar a fouled ball or spool loose. Then the windmill will go back to pumping, provided the sucker rod is not broken.

If the valves can be heard to be working, and still no water is being pumped, it's probably a hole in the water column. About the only way to check this is to try and see if it will hold water. Even if the well is pumping out, with the cylinder drawing air, there will usually be enough water in the water column to surge and bubble. In this case, pull everything out of the well, get an exact well depth measurement, and hope that there is enough room to lower the cylinder or tail pipe to where it will be able to always pick up water.

When — for any reason — you replace all or part of the sucker rod string but do not replace the cylinder, make sure that the plunger valve is either pumping in the original stroke area or that you have changed it to pump completely in a new section of the cylinder. Failure to do this may cause a cup leather to turn back because it travels where the diameter of the cylinder is smaller, having not been previously worn larger by the original stroking of the leathers.

A sand point.

For all windmills, sand is a problem often encountered. It may be partially solved by a sand point used as a strainer, with the screen size depending on the grain size of the sand.

Silt, especially in low yield wells, may present even more of a problem than sand. In such wells pumping will cause fluctuation of the static water level. This in turn causes the water to become discolored with fine silt. The silt will often plug the screen to the extent that water cannot get into the cylinder, or perhaps just enough water is drawn in to create a vacuum.

Caused by pumping more water out than is allowed in through the cylinder, this vacuum may be detected by the sucker rod "bumping" on the down stroke. The "bump" actually follows the stroke downward instead of remaining at the same level with each stroke. I have no idea and no way of finding out just how many cylinders this strange little bump has caused to be replaced. Because my honesty outweighs my better judgment, I will admit to replacing two cylinders before it dawned on me what was causing the problem. (Moral: clean the screen.) And here is a situation where I much prefer ball valves rather than spool valves. The balls are not as likely to become fouled as the spools, and because of their shape and weight they have a better chance of cleaning themselves if they should become dirty.

Another maintenance problem of no little magnitude is tree roots. At least in my area I'll head the list of offending trees with Chinese Elms. (I don't even feel comfortable sitting in the shade of one.) There are also cottonwoods and willows.

When releathering the plunger valve, care should be taken not to tighten the valve to the extent that the cup leathers spread and will not go into the cylinder. If the bottom valve has been pulled during releathering, it is common practice to put on a new cup leather and drop the valve back down the water column. This practice has its drawbacks, however, especially in wells of any depth or in wells where there are rust barnacles in the water column. When the bottom valve has been dropped and is seated, then whatever rust is dislodged when running the sucker rod and plunger valve will often foul the bottom valve. One way to prevent this—to a certain extent—is to screw the bottom valve on loosely, (four or five threads will do), to the plunger valve. Then simply let them both down together. When the bottom valve is seated, unscrew it from the plunger valve and you are in business.

I have never had much success in lubricating cup leathers with cup grease, tallow, bacon grease, or the like. It is my theory that they all cause rust and corrosion to adhere to the cup leathers as they are lowered to the cylinder, thus creating another problem. The best way that I have found to lubricate cup leathers came about more or less as a joke, this by way of my regular windmill salesman. He told me that an old Indian over on the Mescalero reservation had told him to just put eggs in the water column. We laughed about that, and went on to discussing other things that we thought we were so wise about. At breakfast the next morning, with two eggs smiling up at me, I thought about the Indian. Then and there, I decided that the next time I encountered a tight cylinder, I would test the Indian's wisdom. I didn't have long to wait. Two or three days later an old rancher friend called and told me to come out and bring a new cylinder and red rod. He had releathered his valves, but then the windmill began picking up the water column pipe (because of a stuck cylinder) finally breaking the red rod. I went on out with a new cylinder, a new red rod and a half-dozen cheap eggs. I had to cut the lock-chain on the gate in the back side of his pasture where the windmill was, figuring that I would be able to carry out my experiment in secret. No such luck. About the time I had the red rod replaced this rancher drove up and there was my egg carton sitting on a plank that ran through the tower. After he had inquired about my well-being and explained how long that old cylinder had been in the well, his curiosity finally got the best of him. "What in th' hell is in th' egg carton," he asked. "Eggs," I replied. "Well yes, I reckon that figgers, but what for?" When I told him that I planned to see if I could make his cattle roost in the trees, the only response I got was a "humph." It was then that I realized that my entire reputation and professional standing in the "Windmiller's Hall of Fame" was wrapped up in the paper-thin shells of six pullet eggs, and I had a pretty good idea of how Bennie Franklin felt when he flew the kite. Somewhere along the line I had heard the saying "nothing ventured, nothing gained," and besides, I had gone too far to back out now. When I started cracking eggs and dropping them down the water column, my friend got up off the tail gate of his pick-up and said, "I'm almost eighty years old and I've worked on windmills all my life, but I never saw anything like what you're doing!" I assured him that that made two of us, and if it worked we would tell everyone and if it didn't we'd never tell a soul. I waited for three or four minutes until I figured the egg yolk had time to sink down to the cylinder, then climbed the tower to turn the wheel. Sure enough, like the rancher had said, up came the pipe along with the sucker rod. I set the pipe back down and gave the wheel a few quick jerks, gaining about a half an inch with each jerk until I had a full stroke. After eight or ten strokes, I put the mill into the wind, and it went to pumping. While I was gathering up my tools, the rancher got in his pickup and left. About fifty yards down the road he stopped and backed up to where I was, stuck his head out the window, and hollered: "Say, Rooster, if you'll weld that gate-chain back together I'll send yuh a key for th' lock."

Now then, I do not intend to leave the impression that even a better grade of eggs will work every time in a situation like I have just described, but if you don't mind oatmeal for breakfast, then it's worth a try.

Pulling sucker rod and releathering valves is ordinarily a topic that deserves very little discussion. Anyone capable of putting his pants on the right way, and who has the right tools, ought to be able to do the job. There are times, however, when ordinary is not the case. Let's say that there are problems, like the well has caved in and the cylinder is full of mud, or perhaps the plunger valve was lowered and screwed on to the bottom valve, or maybe a cup leather has turned back and the valves won't come out of the cylinder. With any of these, the first thought that usually comes to mind is more lift power, and right behind this brain wave comes the use of hydraulic jacks. This is kind of like sending a five-year-old boy rabbit hunting with a 30-30 rifle; a lot of things can happen when you resort to "more power." Probably ninety percent of the time when you put a jack between the water column and the sucker rod you'll do one of two things: either you will pull the valves or you will simply break the sucker rod in two. This is no big deal as everything is still intact. But that remaining ten percent is something else again. Let's say that we have anywhere from three to six-hundred feet of water column and sucker rod in a well. And as is usually the case — to complicate matters — let's say the water column is made up of standard pipe and collars. The first problem is that when a water column this length is full of water, it's a pretty fair load for those standard collars that are up toward the top of the water column. Now in this case if we have jacks between the water column and the sucker rod, and because of the weight of the water column there is anywhere from eight to eighteen inches of stretch in the sucker rod, we have a potential problem. If the rod breaks, especially near the top of the string, we are going to have the thrust of the stretch plus the weight of the rod going straight down the well. What collars didn't break will more than likely be broken when that water column weight hits the bottom of the well. Fishing pipe that is intact and open out of the well is not usually too much of a problem. But when you have twenty or thirty joints of dog-legged pipe wadded up in a well, some with sucker rod still inside and some without the rod, then this spells trouble with a capital "T."

In a situation where the valves are stuck, it would be much better to try and hammer them out of the cylinder with the sucker rod jar as described in Chapter II. Failing this, pull the water column and sucker rod together and salvage whatever pipe and rod you can. At least, this may save the rod and the water column.

I believe that one of the most common causes of stuck cylinder valves and even stuck water columns stems from a hole in the water column in uncased or partially cased wells. A hole the size of a pin head or match stem will easily allow the windmill to fill the water column as if nothing were wrong. In time, the force of the water streaming out of this hole will dig a large hole out of the wall of the well. When debris from this erosion settles around the cylinder or the water column, they may well become stuck. So, any time a water column is not holding in an uncased well, it should be pulled and checked carefully for holes.

Starting a water column out of a well is another operation where care must be exercised. This is a good time to remember that old proverb, "Haste makes waste." I have already mentioned that whoever is operating the power unit — whether it be winch, truck, blocks, or whatever — should be the person with the most experience in this type of work. When starting the load, he has to "get the feel of it." If the water column is mudded-in and perhaps stuck, this is a situation where jacks can be used to help loosen it. If the water column can be raised at all, even an inch or two, then let it back down and repeat the procedure, trying to gain a little each time. This action will usually loosen the mud enough to enable you to eventually get the water column and cylinder out of the well. Even if the water column is free, it's wise not to get in a hurry. A collar could easily hang on a ledge or a casing joint, or a rock could fall out of the well wall and lodge against a collar. If you are in a hurry when any of these things happen, you'll no doubt break something.

Frequently wells of low yield will weaken or perhaps fail. This doesn't always happen in a drought. It can also occur following a long wet season when creeks are full and springs are gushing full blast. When such a failure occurs during a drought, it is generally assumed that the water course has dried up. In new wells, however, drilled in rock formations where the water comes in through small seams or cracks, and when the well was drilled with a rotary rig or down-the-hole hammers, the seams and cracks may easily be sealed with the cuttings from the drilling rig.

I am of the opinion that when the well weakens during wet times, it is because the surge of water through the cracks and seams has enough pressure to move debris around. This debris will become lodged in and seal other water courses. If the water is there, but sealed off for whatever the reason, many times the flow may be restored by using dry ice.

Usually about fifty to one hundred pounds of dry ice will do the job. The amount depends, though, on the depth of the well and how much ice has melted before it reaches the bottom. For maximum efficiency, the ice should be broken into pieces small enough to go down the well. Get all or as much of it as possible in the well before it starts blowing back out the well head. In wells with only two or three lengths of casing, no attempt should be made to plug the well, as this could put enough pressure on the whole thing to blow the casing right out the top. If the casing is to be plugged, it can be done with a green tree limb or small tree trunk whittled to fit and driven in by two men and their sledge hammers. This has to be done fast, and be sure not to sit down on the plug to rest. No attempt should be made to install the water column for three or four hours, as it may freeze because of the still intense cold.

What happens down in the well is that the ice causes the water to boil and churn. This strong action will usually loosen the debris in the cracks and seams, and restore the flow of water. At least, that's what it's supposed to do. If there is any water in the well at all, this will work. But if the well is already completely dry, then this dry ice treatment will not work at all. If you are in doubt, you may as well try it, because this fix can't hurt — not like dynamite.

The last thing I need to mention is fishing. Fishing for sucker rod depends a lot on why. Rod that has been broken or has become unscrewed does not present too much of a problem as the water column may always be pulled to get the rod. Heavy strings of rod that have been dropped may be an entirely different matter, especially if the water column does not have a lot of water in it. The rod may knock the cylinder off and if the water column is pulled leaving the sucker rod in the well, the latter is much more difficult to fish. If the pump rod swivel in the mill is not working properly, and there is no guide in the tower to keep the sucker rod from turning, the rotation of the windmill on the turntable could possibly unscrew a sucker rod joint down in the well. In a situation such as this, many times, with little fishing and a lot of luck, the joint may be screwed back together. When and if this has been accomplished the sucker rod string should be pulled, at least to the joint that came loose, and that joint retightened. Failing in this endeavor, or when fishing for rod that has been broken, your best bet is an over-shot. The over-shot can be made out of a piece of pipe small enough to go inside the water column, large enough to go over the sucker rod, and long enough to go to a joint. A slot twelve or fourteen inches long and about one-and-one-fourth inches wide is cutout near the bottom of the pipe. A piece of truck spring is welded at the bottom of the slot. Heat this spring and force it to the inside to fit the rod that is being fished for. An over-shot may also be made large enough to fish for the outside of pipe collars. The best way to use this tool is to weld a sucker rod joint to the top and run it with sucker rod.

When fishing for the inside of pipe—two, two-and-one-half, three, or four inches — one of the first things to take into consideration is the weight of the pipe that was dropped. Make sure that you are rigged up to handle more than that much weight, because pipe may be dog-legged or stuck in the mud, or possibly both. The best tool that I have found for fishing for the inside of a pipe is the slip-spear. It consist of a three- or four-foot tapered bar with a threaded point that screws on the bottom of the bar. The bar is slotted on two sides for the slips to run in. The two corrugated slips are four or five inches in length. When the bar is inserted inside the pipe the slips slide up the taper of the bar until the bar will fit inside the pipe. When the spear is pulled back up the slips tighten against the wall of the pipe and the harder the pull the tighter they get. It's a good way to get pipe.

APPENDIX

WINDMILL TROUBLE SHOOTING CHARTS

NOTE: Refer to diagrams page 47 and 48.
PROBLEM AND CAUSE — AND REMEDY

1. **MILL FURLS OUT, BUT WHEEL CONTINUES TO TURN SLOWLY**
   a. Excess wear in upper furl lever connections.
      REMEDY: *Furl arms #528 need to be replaced with #528¾ which are extra long to compensate for wear. A lot of #690 brakes are installed but do not correct problem, as not enough pressure is being applied due to wear in linkage.*
   b. Brake band #690
      REMEDY: *No brake lining is used on band, depends on metal contact. Replace if worn excessively.*
   c. Furl lever linkage worn in rivets and connections of #609.
      REMEDY: *Replace furl lever #609 on 4 post tower or #670 if mill is on 3 post tower.*

2. **MILL JERKS WHILE OPERATING**
   a. Cylinder leathers may have swelled and binding in cylinder barrel.
      REMEDY: *Replace leathers.*
   b. Cylinder barrel may have been crushed due to pipe wrench being used on barrel.
      REMEDY:- *ever use pipe wrench on cylinder barrel, especially a brass barrel. Replace cylinder if crused or scored.*
   c. Pump rod coupling may be catching at pipe joint.
      REMEDY: *Find problem connection and correct.*
   d. Stuffing box nut may be too tight and binding.
      REMEDY: *Loosen and repack stuffing box. If rod in stuffing box is pitted or scored, replace rod.*

106

3. **MILL KNOCKS AT BOTTOM OF STROKE**
   a. Yoke #608 is hitting top of mastpipe locknut #578.
      REMEDY: *Motor position has to be raised.*
   b. Turntable washers worn and let crated motor drop.
      REMEDY: *Install 2, 3 or 4 #521 split washers on mastpipe at turntable and greasepack.*
   c. Plunger in cylinder hitting cage of lower check valve in cylinder.
      REMEDY: *Rod adjustment necessary. Shorten pump rod. (Rod will sometimes stretch a little after installation, especially on deep setting.)*
   d. Stuffing box may be too tight.
      REMEDY: *Loosen and repack gland.*
   e. Pitmen arms may be worn excessively.
      REMEDY: *Replace or rebabbitt.*
   f. May have changed partially to short stroke.
      REMEDY: *Must change lower pitmen to enter boss on large gears and in bottom hole at top of pitmen arms.*

4. **PITMEN ARMS KEEP BREAKING**
   a. Large gears misaligned. Causes pitmen arms to operate in a bind.
      REMEDY: *Remove gears #705 and line up boss on each gear so each is in alignment with each other.*
   b. Bearing #752 between large gears worn excessively.
      REMEDY: *Replace #752 bearing and #720 shaft if necessary.*
   c. Replaced only one pitmen or one gear in field.
      REMEDY: *Always replace both pitmen arms or both gears, so load is distributed evenly.*
   d. Mill may be overloaded.
      REMEDY: *Check recommended table and cylinder size. Setting mill on short stroke may alleviate condition.*
   e. Mill may be loaded too much on downstroke due to stuffing box or counter-balance. Pump pole in tower will usually bend or buckle a little.
      REMEDY: *Check stuffing box or correct counter-balance.*

5. **OIL LEAKING DOWN MAST PIPE**
   a. Too much oil in gear case, and runs into hole for oiling turntable.
      REMEDY: *Do not overfill. Check helmet #560 for bullet hole where water could get into gear case and float oil out.*

b.   Lockwasher #579 for top of mastpipe may be off to one side and rubbing on inside of large gears and picking up oil.
REMEDY: *Be sure lockwasher #579 is centered and locknut is tight.*

6. OIL LEAKING AROUND HUB AND ONTO WHEEL

a.   Oil collector #520 plugged and does not pick up oil from oil pocket inside of hub.
REMEDY: *Remove #520 oil collector and clean and flush out oil passage with kerosene.*

b.   Using too heavy a grade of oil which won't flow and plugs passages.
REMEDY: *Drain and flush system with kerosene. Refill with correct weight oil.*

c.   #730 shaft bearing may be worn excessively.
REMEDY: *Replace if necessary.*

d.   May be using detergent type of oil.
REMEDY: *Use a non-detergent type of oil.*

WINDMILL
MAINTENANCE AND INSPECTION

NOTE: **Refer to diagrams page 47 and 48.**

1. TOWER
a.   Check to be sure tower is plumb and true.
b.   Check girts, angle braces. Are they bent or missing?
c.   Check all bolts to be sure they are tight.
d.   Check condition of wood pump pole and pole guides, splices.

2. WHEEL CONDITION
a.   Are all bolts tight or missing in inner and outer bands?
b.   Check rivets or bolts in sail ribs.
c.   Check hub and wheel arms.
d.   Does wheel turn true and free?

3. WINDMILL HELMET IN POSITION
a.   Check for damage, such as bullet holes.
b.   Is helmet positioned correctly or bent out of shape?

4. FURLING DEVICE
a.   Check linkage connections for wear or missing rivets.
b.   Examine furl arms, tailbone casting, brake lever and linkage.

5. CHECK TAILBONE, VANE
a.   Is tailbone bent or pivot bolt nut on securely?
b.   Check vane spring for wear at connections.

6. **SECURE WHEEL SO IT CANNOT TURN NOR CAN MILL SWING AROUND**
   a. Check wood platform for rot, oil which would make it slick.
   b. Check area for bees, wasps, or other insects and eliminate before working on tower, for your own personal safety.

7. **REMOVE NUT SECURING HELMET AND REMOVE HELMET**
   a. Check oil level — is oil clean?
   b. Examine large gears for wear, broken teeth.
   c. Check pitmen arms #686 for wear.
      (1) Is #622 pin and cotter in position?
      (2) Check for side play of pitmen arms.
      (3) Check upper shaft for guide wheel and yoke.
   d. Is #751 bearing bar securing gears properly?
   e. Try moving large gears sideways for wear in #752 bearing.
   f. Check oil ring.
      (1) Is it round or egg shaped — must be round to roll on large gears at bottom of stroke.
      (2) Should hit 3 to 4 teeth and roll.
      (3) Provides lubrication to upper guide wheel and shaft.
   g. Is spring #517 holding #718 spout washer in position?
      (1) Spout washer must rub small pinion gear.

8. **CHANGING OIL AND LUBRICATION**
   a. Remove the drain plug at the front near the bottom of case.
      (1) Catch oil in container so that it cannot get into well nor onto platform where it could become slick and dangerous.
   b. Look for grit, sand in bottom of gear case.
   c. Turn wheel so pipe plug in wheel hub is on top.
      (1) Remove pipe plug in hub and flush out the oil collector and passages with kerosene.
      (2) It may be necessary to remove the #520 oil collector to flush return passage with kerosene.
      (3) Replace #520 oil collector and tighten. Oil collector must rub the inside of the hub to pick up oil and return this oil to gear case.
      (4) If installing new #520 oil collector:
         a. Screw in securely.
         b. Turn wheel slowly by hand about 15 degrees and then reverse wheel and turn in other direction. This will seat the oil collector and it should drag or rub on the inside of the hub.
         c. Check again to see that oil collector is tight.
      (5) Replace pipe plug in hub housing.
   d. Replace drain plug in case and tighten securely.
   e. Refill the gear case with oil.

 (1) Do not overfill with oil, as too much oil may cause overflow at the wheel-hub.

 (2) DO NOT USE HEAVY OIL. Use a light oil that flows freely to all working parts. Heavy oil may clog the oil passages, causing extra wear on some of the working parts, due to lack of lubrication.

 (3) AERMOTOR OIL. Non-detergent oil. (Detergent oil will adhere to shaft and get by the oil collector. Not like a car engine where excessive heat tends to break down the oil, and no carbon build-up, so non-detergent oil is recommended.)

f. Oil the turntable.

 (1) Through the hole in the main frame near the top of the supporting pipe.

g. Oil the furl ring.

 (1) Some remove the plug and insert a zerk fitting so they can grease it.

h. Grease pack the pump pole swivel.

NOTE: In some sections of the country, especially where there are a lot of dust storms, it may be advisable not to oil or grease external parts, such as furl ring, tailbone pivot bolt, brake lever pivot, etc., as sand, dirt becomes imbedded in grease or oil and causes further or faster wear.

9. CHECK #608 YOKE AT BOTTOM OF STROKE

a. Turn mill over slowly and at bottom of stroke be sure bottom side of #608 yoke is not striking mast pipe locknut.

 (1) If it hits bottom of stroke or is close to striking

 (2) Needs turntable washer #521 (split washer) to raise motor position.

  a. Necessary to lift up the motor and block it up.

  b. Grease pack the #521 split washer before installing.

  c. Be sure to tighten locknut #578 when completed.

10. INSTALL HELMET #560

a. Very important to position correctly, to keep out dust, dirt and rain.

11. CHECK FOR LEAKS AT DRAIN PLUG

12. FIELD SERVICE AND REPAIR

a. On site or location.

 (1) Most common parts requiring replacement:

  a. Pump rod #171 and #610 pin for yoke. Reciprocating action — jerking due to wear. Radial action — due to wind directional changes.

  b. Pitman arms #686. Always replace both pitmen — balanced load.

  d. Vane spring #28.

  e. Shaft #522 for guide wheel and yoke.